SpringerBriefs in Earth System Sciences

Series editors

Jorge Rabassa, Ushuaia, Argentina
Gerrit Lohmann, Bremen, Germany
Justus Notholt, Bremen, Germany
Lawrence A. Mysak, Montreal, Canada
Vikram Unnithan, Bremen, Germany

More information about this series at http://www.springer.com/series/10032

Manuel E. Pardo Echarte
Osvaldo Rodríguez Morán

Unconventional Methods for Oil & Gas Exploration in Cuba

The Redox Complex

 Springer

Manuel E. Pardo Echarte
Scientific-Research Unit Exploration
Centro de Investigaciones del Petróleo
El Cerro, La Habana
Cuba

Osvaldo Rodríguez Morán
Scientific-Research Unit Exploration
Centro de Investigaciones del Petróleo
El Cerro, La Habana
Cuba

There are instances where we have been unable to trace or contact the copyright holders. If notified the publisher will be pleased to rectify any errors or omissions at the earliest opportunity.

ISSN 2191-589X   ISSN 2191-5903   (electronic)
SpringerBriefs in Earth System Sciences
ISBN 978-3-319-28015-8   ISBN 978-3-319-28017-2   (eBook)
DOI 10.1007/978-3-319-28017-2

Library of Congress Control Number: 2015958320

Printed on acid-free paper

This Springer imprint is published by SpringerNature
The registered company is Springer International Publishing AG Switzerland

Foreword

As a rule worldwide, the first, and many of the most important oil discoveries have been made following the method of drilling near springs, asphalt deposits or oil seeps, based on the assumption that some hydrocarbons of an oil-gas accumulation migrate to the surface vertically. Special attention deserves the known microleakages or microseeps given its relationship with well-preserved hydrocarbon accumulations at depth. The probable mechanisms consider the vertical rise of gas bubbles colloidal size through the interconnected system of microfractures, cracks and bedding planes, all filled with water in the subsoil. Floating leads these bubbles to the surface almost vertically. As the light hydrocarbon rise, the bacterial oxidation produced as by—products carbon dioxide acid and hydrogen sulfide. For its part, the carbonic acid reacts with the clay, destroying minerals, while creating secondary carbonate mineralization and silicification. Both result in making the surrounding surface materials denser and erosion resistant, with an effect on the increase of seismic velocity on the accumulation and the formation of erosional topographic maximum. Meanwhile, the destruction of clay minerals by the acids mentioned releases potassium from rocks which is leached by groundwater and surface water. However, the thorium remains relatively stable in its original distribution within the insoluble heavy minerals. Thus, it becomes evident that the ratio of attributes K/eTh is informative (for their minimum values) of the environment affected by the process described above. Additionally, this relationship offers the opportunity to eliminate a number of undesirable effects on spectrometric measurements (influence of lithology, moisture, vegetation and measurement geometry). On the role of sulfhydric, its own presence conditions the formation of a column of reducing environment (minimum of Redox Potential) on accumulation. This reducing environment favors, in turn, converting the non-magnetic iron minerals in more stable magnetic diagenetic varieties as magnetite, maghemite, pyrrhotite and griegita all responsible for the increase (maximum) of Magnetic Susceptibility of rocks and soils on the occurrence, which explains the observed inverse correlation between attributes (minimum of Redox Potential and maximum of Magnetic Susceptibility) and justifies the integration of methods.

Oil-gas unconventional geophysical-geochemical exploration techniques, among which is the *Redox Complex*, are indicative of physical—chemical processes and/or environmental changes taking place in the upper section directly on the occurrences conditioned by diffusion halos of light hydrocarbons and other satellites elements (metal ions) that reach the surface. These techniques are used as a complement to conventional prospecting complex with the purpose of assisting in the reduction of areas and/or selection of the most favorable geological testing targets, thereby increasing the effectiveness of geological investigations.

The antecedent in the use of the referred techniques back to the decades of the forties and fifties, where aerial radiometry to map hydrocarbon deposits (minimum total counting on the areas of production) was used. In the seventies, the methods airborne gammaspectrometric high sensitivity (minimum potassium), aeromagnetic (micromagnetic anomalies) and Magnetic Susceptibility of soils (maximum) applied to oil exploration were developed. In Cuba, the application of these techniques has its antecedent in the work of Alfonso Roche and Pardo (1993) in which the possibility of remote mapping potentially producing areas, revealed by lows of K/eTh relationship, and verify the oil-gas nature of these anomalies, specifying their limits on land by the conjugate expression of minimum of Redox Potential and maximum of Magnetic Susceptibility of soils. Industrial accumulations of hydrocarbons studied (Varadero, Cantel and Pina oil fields), revealed by lows of K/eTh relationship to values lower than 0.1 are expressed by minimum of Redox Potential in soils with amplitude between 30 and 80 mV, whose epicenters match the apical part of the structures, being reflected the direction of its sinking by the lower flank gradient. For antiforms, the Redox anomaly has a maximum relative over its axial part, coinciding the epicenters approximately with the limit water/oil. Soil magnetic susceptibility, as a rule, is less informative, even though, in some cases their maximum may be diagnostic of the production structure position.

In any country it is essential the ongoing assessment of hydrocarbon potential for the renewal of the exploratory oil policy and strategy of its economic development. This is supported by the increased acquisition of new geological, geophysical, geochemical and other data and the constant refinement of the criteria and methods of oil exploration, which provide a plenty of important information. In recent decades unconventional exploration methods have come to harmoniously complement traditional methods, by virtue of improving outcomes for oil exploration.

The *Redox Complex* (Redox Potential, Magnetic Susceptibility, Spectral Reflectance and Soil Geochemistry) is a complex of unconventional exploration techniques, used for indirect detection and evaluation of various objects of metal nature, which is based on the *Geochemical Principle of Vertical Migration of Metal Ions*. It is a genuine Cuban combination of techniques. It has been applied in various fields (Oil, Metal Minerals, Environment and Archaeology) since 1997, initially rudimentary and subsequently incorporating new theoretical—practical elements in their work. So, the *Redox Complex* has been used to elucidate the structure and nature of hydrocarbon prospects in Cuba, mainly in the Northern Fringe of heavy crude and Central Basin. This complex of unconventional exploration techniques has well identified its possibilities and limitations, making it more objective and credible results. Although the available applications are still statistically insufficient as experimental foundation of empirical regularities formulated mathematically, however, they allow, a priori, envision a better future for this complex of exploration methods.

The reader will find in this essay, the most important insights and results related to the *Redox Complex*. The contribution of its principal author has been decisive in obtaining them. His expert status in the acquisition and interpretation of the data in this complex of unconventional exploration techniques has led to it for an important place in the work of oil exploration in Cuba.

Dr. Evelio Linares Cala
Head of Regional Geology and Geological Survey Department
Oil Research Center

Preface

The *Redox Complex* (Redox Potential, Magnetic Susceptibility, Spectral Reflectance and Soil Geochemistry) is a complex of unconventional geophysical-geochemical exploration techniques, used for indirect detection, characterization and evaluation of various metal targets, which is based on the *Geochemical Principle of Metal Ions Vertical Migration*. This procedure offers information on the shallow terrain modifications, over any metallic target, controlled, under the referred Geochemical Principle, through their composition, grade, geometric features and lying characteristics. Its final product consists of: a map with the cartography of the vertical projection of the metal target; a central section with its geometrization, together with the behavior of the attributes that characterize it; the parameters defining the nature and quality of metallic source and a resource estimate. This complex is successfully applied in various fields: oil and gas and metal ores exploration; studies of oil and metal contaminants in soils; and the search for metallic archaeological burials. The use of these techniques is intended to complement the conventional prospecting complex with the purposes of reducing areas and/or the selection of the most favorable targets, resulting in an increase in economical-geological effectiveness of investigations. The *Redox Complex* is implemented without physical or chemical damage to the environment.

The combined application of Redox Potential and Magnetic Susceptibility of soils to geological prospecting, which is the *Redox Complex* antecedent, is determined by the possibility of detecting the reducing column that is formed directly on hydrocarbon and metal occurrences and reaches the upper part of the section. Within this column, the conversion of non-magnetic iron minerals to more stable magnetic diagenetic varieties is favored, which explains the observed inverse correlation between the attributes and justifies integration of methods.

It is outlined the general features characterizing the processes of metal mobilization, vertical transport to the surface, and the resultant accumulation of these vertically transported metals in surficial media from hydrocarbon deposits and buried ore bodies: microbial activity and water-rock reactions with gas (hydrocarbons, N, CO_2, H_2, and others) generation during target oxidation; ascending reduced gas microbubbles (colloidal size) with reduced metal ions attached, which results in 'reducing chimneys' reaching the surface; barometric pumping and capillary rise moving upward ions and sub-micron metal particulates, into the unsaturated zone; redistribution of ions in the near-surface environment by downward-percolating groundwater (after rainfall) as well as by the upward effects of evaporation and capillary rise, all of which explain the soil metal accumulations in a very shallow (10–30 cm) 'metal accretion zone'.

It is designed a database and applications system (*Redox System*) to solve in a quick and reliable way, all the storage processes, reports, graphics and the corresponding interpretations of the *Redox Complex*. To carry out the design of the different kinds of applications, which answer to the qualitative and quantitative data interpretation, it is model, mathematically, the expert's experience. For this objective, the methods of the Knowledge Engineering and the techniques of the UML (Unified Model Language) diagrams are used. The *Redox System* transcends the limits of a simple calculation and storage program, which it can be used by other specialists without being expert in the topic to obtain results on interpretation. A User Manual of the Redox System is drawn up.

The illustrations of some of the applications of geophysical-geochemical unconventional methods for oil exploration in Cuba are presented. These consider the regions of Habana-Matanzas (Varadero Oil Field, Cantel Oil Field and Madruga Prospect) and Ciego de Ávila (Pina Oil Field, Cristales Oil Field, Jatibonico Oil Field, Jatibonico Oeste Prospect and Cacahual Prospect). The methods considered contemplated, in some cases, airborne gamma spectrometry (K/eTh ratio) and reduced Redox Potential (ORP) and, in others, the *Redox Complex* with reduced or standard attributes. In all cases, the anomalous complex of interest corresponds to the correlation of minimum K/eTh ratio, minimum ORP and, in the case of *Redox Complex*, Magnetic Susceptibility highs with ORP lows, Spectral Reflectance lows and maximum Content of Chemical Elements.

Contents

Chapter 1
The *Redox Complex*: Methodological and Theoretical–Empirical Considerations

Abstract The growing need for geological prospecting as well as environmental and archeological studies asks for an effective increment of conventional research at reduced costs. With these purposes unconventional geochemical techniques of indirect detection, employed with complementary character, such as *Redox Complex* (Redox potential, magnetic susceptibility, spectral reflectance, and soil geochemistry), are identified. This procedure offers information on the shallow terrain modifications, over any metallic target, controlled, under the *Geochemical Principle of the Vertical Migration of Metallic Ions*, through their composition, grade, geometric features, and lying characteristics. For this, it is effective in its location, characterization, and evaluation, enabling the optimization in decision making and raising the success rate of the drillings and/or exploratory excavations. Disclosed is a new unconventional geophysical–geochemical exploration tool for indirect detection, characterization, and evaluation of targets of metallic nature, in which final product consists of a map with the mapping of the vertical projection of the metal target; a central section with its geometrization, together with the behavior of the attributes that characterize it; and the parameters defining the nature and quality of metallic source and a resource estimate. This scientific–technological innovation is at present in the generalization stage for its further application in the different spheres.

Keywords Mobile metal ions (MMI) · Unconventional geophysics-geochemical exploration · Metallic sources · Resource estimate · Buried ore bodies · Hydrocarbon deposits

1.1 Introduction

The need to raise the geologic-economical effectiveness of minimum programs by conventional methods in geological prospecting leads to complementation by unconventional geophysical–geochemical techniques such as *Redox Complex*

© The Author(s) 2016
M.E. Pardo Echarte and O. Rodríguez Morán, *Unconventional Methods for Oil & Gas Exploration in Cuba*, SpringerBriefs in Earth System Sciences, DOI 10.1007/978-3-319-28017-2_1

1

(Redox potential, magnetic susceptibility, spectral reflectance, and soil geochemistry), which provides information on environmental changes taking place in the upper part of the section, directly on a metal target, under the *Geochemical Principle of Vertical Migration of Metal Ions* toward the ground. As a rule, these changes are controlled by the peculiarities of composition, grade, geometric, and lying features of the metal object, so this complex is particularly effective in its location, characterization, and evaluation, enabling the optimization of decision making and raising the success rate of the drilling and/or exploratory excavation.

This research is supported by a precedent work: "*Method for Measuring Soil Redox Potential and its Combined Application with Kappametry for the purposes of the Geological Prospecting*" which was the subject of the Certificate of Author Invention No. 22 635, given by Resolution No. 475/2000 OCPI, and the Legal Facultative Deposit Certification of Protected Works (Copyright 1589–2005) for the "*Redox System.*"

This research has three main objectives: increasing the number of attributes to be measured in order to improve the multiparameter characterization of metallic targets and thereby expanding application areas of the complex; to establish the methodology of the various observations and quality controls; and the empirically achievement of a mathematical formulation of such regularities linking the anomaly response of the measured attributes to the characteristics of lying, composition, and rough quality of the metal sources in order to establish the quantitative interpretation process.

1.2 Principle of Vertical Migration of Metal Ions

The techniques of selective extraction of metallic elements linked to the *Principle of Vertical Migration of Metal Ions* have become popular in mineral exploration since the 1980s and 1990s, may cite, as examples, the techniques CHIM (in Russian, Metal Partial Removal) (Moon and Khan 1990) and MMI (Mobile Metal Ion) (Birell 1996; Mann 1997).

At the CHIM (Moon and Khan 1990) technique, a series of special metal-selective electrodes are placed on the ground at stations with regular spacing. A direct current is passed through each electrode, in parallel, with the return current passing through a distant electrode (at infinity). An electrolyte in each electrode is periodically sampled and analyzed in the field for specific metals of interest. The peculiarity of this method is that it selects and focuses mobile metal ions which occur in concentrations (in ppb) below the analytical limits of accuracy (ppm) available then. As a transport mechanism of the "fast" ions the method considers some source gas or "Geo-gas."

In the MMI (Birell 1996) technique the subtle metal ion concentrations loosely adhered to the surfaces of soil particles are mainly measured, from a source of deep or buried mineralization, using mass spectroscopy by inductively coupled plasma (ICPMS). The popularity of this technique results from the fact that the selective extraction of metal elements (detached in solution by the action of weak digesters which minimize attack of the matrix material) from a mineral ion source significantly deeper or covered by thick exotic deposits exhibits narrow anomalies that are located close to the occurrence apical projection and with higher contrast anomaly/background than those for conventional digestion (where strong acid digesters dissolve most of the chemical matrixes of the soil) whose anomalies are broader than those involved in the former. The conditions under which these anomalies occur reveal a high mobility (speed) of the metal ions, hence also its name as "fast" moving metal ions. In the MMI technique (Mann et al. 2005), empirical observations suggest that anomalies are preferably located in a range between 10 and 25 cm below the soil interface (below the organic layer), regardless the horizon to which this depth corresponds, and comprising elements contained in the ore which are directly located on the source of mineralization. Laboratory experiments suggest that both the capillary rise and the evaporation play an important role in determining the position of the metal ions into the soil profile: the transpiration root zone is also involved in the deposition/adsorption of solute within the evapotranspiration zone. The effects of rainwater percolation and upward forces of capillary nature are considered in a model explaining many of the features of the position of the metal ions in soils. Laboratory modeling also suggests that the convection probably due to the heat produced by the oxidation of the mineral occurrence may, in some cases, provide a mechanism for the rapid rise of the ions below the water table. Another alternative source of ascent can be the hydromorphic transport of metal ions from weathering processes. Research on this technique has been mainly focused on the mechanisms of anomaly formation (transport mechanisms and deposition of metal ions), and this is considered as an important exercise in fundamental science.

According to Smee (2003), the main uncertainty in the selective extraction techniques is the lack of a solid understanding of geochemical transport processes which could lead to interpretable patterns of elements. Such an understanding is crucial to produce predictive results from a particular mineralized target. Additionally, without this knowledge it is not possible to select the most appropriate method of selective extraction for each geological and climatic environment, neither to choose the most revealing interpretation method. Thus, according to Hamilton (2000), transport processes resulting from chemical weathering of mineralization involves the dispersion of elements under gradients of one kind or another, including chemical, temperature, piezometric (both gaseous and aqueous), and electrochemical. These mechanisms have been invoked as a component of numerous transport and dispersion concepts that have been used by researchers to explain the geochemical anomalies in soils over the mineralization. The models used can be grouped into those that rely on (1) diffusion; (2) groundwater advective transport; (3) gas transport; and (4) the electrochemical transport. Combinations of these models are also possible. According to Kelley et al. (2004), it is also incorporated into the above types, the transport facilitated by biological processes. Of the above-mentioned mechanisms, one of the most relevant and popular, especially to explain the dispersion in indoor environments with a thick exotic young (transported) cover, is the electrochemical transport, which is considered the primary mechanism. This is thought as able to operate even in other environments, but its predominance as the transport mechanism is less certain. Though the models of electrochemical transport reported in literature can sometimes be somewhat controversial, most are based on the same principle: the electron conduction upward along mineralization from chemically reduced deeper areas results in anomalous electrochemical gradients in both the surrounding host rock and the overlying cover; and the mass and charge movement in the form of ions is to develop geochemical anomalies in the cover. By its similarities to the theoretical–empirical views of the *Redox Complex* development it is briefly described, according to the essence of the electrochemical model developed by Hamilton (1998a, b, 2000); Hamilton et al. 2004: At the time of cover deposition (exotic cover), a strong vertical redox gradient is established just above the reduced conductor (ore body) in the bedrock and the oxidizing surface, along which ions have a tendency to move. The outward movement of reduced ions such as HS^-, Fe^{2+}, or $S_2O_3^{2-}$, at the expense of consumption of ionic oxide species that will meet them, which are very limited below the phreatic zone, results in the upward migration of a reduced front from mineralization. Once the front reaches the water table such a process is interrupted due to the predominance of oxidants and a reducing column will have developed in the groundwater saturated cover on mineralization. The aforementioned gradient in the oxidation–reduction potential (ORP) is postulated as the drive mechanism for the redistribution of elements in the cover, explaining the origin of the various geochemical anomalies observed on mineralization, as well as an alternative to the explanation of "reducing chimneys" observed on the occurrence of hydrocarbons. According to Hamilton et al. (2004), studies on electrochemical processes, which result in surface geochemical anomalies, are still in their early stages. Several geochemical processes and seemingly unrelated phenomena occurring on chemically reduced geological features can be tested whether are directly or indirectly linked to the electrochemical processes. Although electro-geochemical techniques have enormous potential, much additional work is still required to ensure reliable and easily interpretable data. Therefore, even if there is abundant literature on the application of spontaneous potential for exploration of various metal targets, only very few works have studied the variations of ORP, among other things, also the problems inherent measuring in surface condition in which, besides being complicated, confronted reproducibility problems.

The *Redox Complex* is not indeed an exception to the conclusions outlined for electro-geochemical exploration techniques. However, whatever the mechanisms of transport operating, the regularity in the functional relationship of ORP and surface concentration of chemical elements from the source, with depth and characteristics of it, seems to correspond, according to empirical observations (Pardo 2004) to an exponential model, rather than a logarithmic model, as would be expected if the process were governed only by the principles of diffusion.

1.3   Possibilities and Limitations of *Redox Complex* in Different Areas of Application

Although applications still available are statistically insufficient as an experimental foundation of empirical regularities formulated mathematically, however, they allow, in principle, the establishment of a set of possibilities and limitations of *Redox Complex* for each area of application:

1.3.1   Hydrocarbon Exploration

(a) Possibilities:

- Recognition and detailed mapping of the occurrence vertical projection and estimation, in an indirect and approximate way, of the probable quality of the oil.
- Rough assessment of the oil occurrence depth and the possible structural trap type.
- Approximate volumetric estimation of the oil occurrence from the detailed mapping results.

(b) Limitations:

- Existence of several overlapping hydrocarbon levels (includes gas caps).
- Failure to seal or it is very fractured.
- The presence of different kinds of reducing surface zones.
- Prevalence of lithological response from the magnetic susceptibility and the spectral reflectance.
- Recently transported cover (less than 10 years).

1.3.2   Metallic Mineral Exploration

(a) Possibilities:

- Recognition and detailed mapping of the occurrence vertical projection and estimation, approximately, of its composition and quality.
- Approximate determination of the depth, dip, and vertical extent of the main ore body.
- Rough assessment of resources.

(b) Limitations:

- Overlapping of primary ore bodies.
- Different kinds of reducing surface zones.

- Prevalence of lithological response from the magnetic susceptibility and the spectral reflectance.
- Recently transported cover (less than 10 years).

1.3.3   *Environmental Studies*

(a) Possibilities:

- Recognition and detailed cartography of the contaminated area, allowing establishing its nature (surface or subsurface) and grade.
- Rough assessment of the depth of the subsurface contaminant source.

(b) Limitations:

- Different kinds of reducing or oxidizing surface zones of various kinds.
- Prevalence of lithological response from the magnetic susceptibility and the spectral reflectance.
- Recently transported cover (under 10 years).

1.3.4   *Archeological Studies*

(a) Possibilities:

- Planimetric position of metal burials and estimate of its composition.
- Rough assessment of the depth of objects.

(b) Limitations:

- Different kinds of reducing or oxidizing surface zones.
- Prevalence of lithological response from the magnetic susceptibility and the spectral reflectance.
- Recently transported cover (under 10 years).

1.4   Methodology of *Redox Complex*

For the in situ measuring of soil Redox Potential (Uredox) two electrodes connected to digital millivolt meter at high input impedance (sensitivity 0.1 mV) are used: an inert platinum electrode, and the other commercial, nonpolarizable copper electrode used as reference. Both electrodes were being located very close to each other in a vertical position, within a 10–30 cm deep hole. The ionic communication which closes the circuit is ensured through the porous ceramic of the reference electrode. Measurements with the described device have a transient behavior. Potential is determined using a five-reading algorithm with 1 min of interval among them. Measurements can be performed also in the laboratory, using a commercially available platinum–calomel cell, in soil samples, rock, or other solid material sprayed. If measurements are made in more than a line, only 30–50 % of the samples are submitted to the spectral reflectance study, and also to the chemical analysis of metals, being used conventional digestions and ICP Spectroscopy for this latter. The quality of the ORP observations is evaluated using the absolute error in the potential assessment, considering 10 % of repeated measurements from the same hole (or samples) used for routine measurements. The accuracy acceptable for the applications described should not exceed 10–15 mV. The lack of repeatability of measurements on different days is solved by the reduction of the background level along with the repetition of at least five points of the previous profile in cases when observations were not daily completed.

The measurement of the Magnetic Susceptibility (Kappa) is usually performed with a kappa meter KT-5 (sensitivity 1×10^{-5} SI), making seven readers on the floor and walls of the hole (or in samples taken from it, with an approximate weight of 150–200 g) which are averaged. Quality of field observations is then evaluated from the relative error in determining the average value of Kappa, considering 10 % of repeated measurements from the same hole (or samples). The accepted accuracy should not exceed 10–15 %.

Measuring the Spectral Reflectance (RE) is performed with a multispectral radiometer CROPSCAN model MSR5, of five bands (485, 560, 660, 830, and 1650) with readings in the first band on samples with a minimum weight of 150 g, using a special design prepared for this purpose (solar light, samples distributed in a circle of 15–20 cm in diameter on a black surface, and the sensor vertically located 30–40 cm above the sample). Quality of observations is evaluated from the absolute error in the determination of spectral reflectance for the first band, considering 10 % of repeated measurements at the end of the working day. The accepted accuracy should not exceed 0.5 %.

Redox Complex field data are captured, stored, processed, interpreted, and represented by the automated *REDOX SYSTEM* specially developed for these techniques, which models the experience of the senior author in applying them. The *REDOX SYSTEM* is easily linked to other professional systems such as MICROSOFT ACCESS, MICROSOFT EXCEL, SURFER, and ILWIS, for outputs as graphics profile and contours maps. It also facilitates access to images of Landsat and Airborne Geophysical Scenarios. Databases created in MICROSOFT ACCESS are converted to XYZ format for better performance.

1.5   Data Interpretation of *Redox Complex*

Qualitative data interpretation of *Redox Complex* involves identifying the increase in soils of metallic elements 'basic constituent' of reservoir, ore, pollution, or metal objects that migrate from the source, with respect to the local background of chemical element characteristic of the residual soil (weathering crust). These may also be affected by various types of alterations related to the source itself. This is also valid for the measured attributes (Uredox, Kappa, and RE), where the qualitative interpretation considers gradient identification (increases or decreases relative to the local background), of each attribute behavior in front of the anomalous contribution of the metallic elements basic constituents from the source. Hence, relevant attributes are dealing with reduced or standard with respect to the local background, depending on the normal or lognormal distribution feature as measured for the attribute. Subsequently, an important aspect is the objective and accurate determination of the different background levels, which determine ultimately the accuracy of quantitative estimates; that is why the authors have designed a unique and sophisticated algorithm to determine the background level for the set of attributes considered. Moreover, the background level varies spatially within the limits of the occurrence and its periphery, being higher on the center part (apical, thicker, or enriched) thereof and diminishing gradually outward. Thus, a qualitative criterion for assessing the proximity to one occurrence (or its portion closer to the surface, with greater thickness or enriched) is gradually increasing the background levels of the metals 'basic constituents', or a progressive decline in the level of background ORP (increased reducing environment).

The qualitative interpretation of different *Redox Complex* attributes (reduced or standard) exhibits its own peculiarities, which are summarized as follows:

- Magnetic Susceptibility normalized (standardized) by the local background (K_n) exhibits highs associated to the column of reducing environment on hydrocarbon reservoirs, mineral occurrences, and pollution at the water table level. This is primarily due to that this reducing environment favors the conversion of mineral nonmagnetic iron (hematite, pyrite) in more stable diagenetic magnetic varieties (magnetite, maghemite, pyrrhotite, and griegite). Often, maxima or lows of lithological nature (areas with magnetite alterations, lateritic, and ophiolite soils; carbonate, kaolinite, and siliceous soils, among others) are observed. The amplitude and morphology of the anomalies of interest is a function of characteristics of the mineral occurrence.
- Redox Potential reduced by the local background (U_r) exhibits lows (on reducing columns) associated to hydrocarbon reservoirs, mineral occurrences, and pollution at the water table level, along with areas of surface reducing environment. On light hydrocarbon reservoirs, the potential has a minimum disjointed character. On dumps, oxidation zones of mineral occurrences, superficial metallic contamination and archeological objects very close to the surface, and potential highs are observed. The amplitude and morphology of the anomalies of interest are a function of the occurrence's characteristics. Redox

Potential is not sensitive to topographic and tectonical variations, being the potential background level characteristic of the type of geological setting (volcanic, oceanic, sedimentary, and/or metamorphic). Even though, in Cuba, soils are mostly of residual type, there is a successful experience on a transported (exotic) cover, for Varadero oil field. The major noisy sources known to date are dry swampy areas, accumulations of organic material, and alluvium with high content in metallic elements, which produce intense lows (≥ 100 mV) of steeper and symmetrical gradients (>1.5 mV/m). Some lithological variations (in soil) can also produce anomalies but at lower order.

- Spectral Reflectance reduced by the local background (RE_r) exhibits, as a rule, lows associated to the increased content of metal elements in soil, as a consequence of its darkness. Frequently, minerals of lithological nature, such as quartz, calcite, kaolin, graphite, hematite, and others, are responsible for highs or lows irrelevant for such study. The amplitude and morphology of the anomalies of interest are a function of the occurrence's characteristics.

- The content of 'basic metal constituent' in soil normalized (standardized) by the local background (C_n) exhibits increases on the vertical projection of the hydrocarbon and metallic mineral occurrences (including oxidation zones), as well as on metal and hydrocarbon contamination and archeological metallic objects. The relationship among the various major elements or basic components is a function of the composition of the occurrence, the contamination, or the metal object. The amplitude and morphology of those increments depend on the characteristics of the occurrence. For the surface contamination, grade is directly determinable from the chemical data.

The quantitative interpretation starts from the general concepts of attribute anomalous response range (AARR), the complex anomalous response range (CARR), the singular points (SP), and the points of absence anomalous response (PAAR). Respective definitions are as follows:

- **Attribute anomalous response range (AARR):** This is the interval of the observation points where the attribute anomalous response occurs.
- **Complex anomalous response range (CARR):** The interval of the observation points where correlatable anomalous response in most attributes occurs.
- **Singular points (SP):** It refers to those observation points within the CARR which satisfy the anomalous behavior expected for attributes. These points are used for quantitative estimates.
- **Points of absence anomalous response (PAAR):** These are those observation points outside the CARR where there is no correlation in the expected anomalous response or simply lack the anomalous response (discontinuity in the source). Some of these points are used to establish the limits of geometric constructions of the source and also to estimate the background level.

The process of quantitative interpretation comprises the following steps:

Step 1

It consists of a careful examination of the graphs and data in order to identify AARR, CARR, SP, and PAAR. Special attention should be paid to the applications of minerals and environment/archeology for the identification of the oxidizing areas, as recognized by the existence of, at least, three continuous points with values of $U_r \geq$ background level + 15 mV.

Step 2

In the database all SP of CARR are selected, which is used to establish linear relationships between attributes, whose output values are a_i and b_i, to be applied in quantitative estimates.

Step 3

It deals with the quantitative estimates of the target depth or thickness performed for each SP linked to linear dependencies for attributes, determined by weight, the most likely value. Once all weighted estimates are performed, one proceeds to the geometric construction of the resulting surface which also uses the PAAR.

Step 4

This step allows performing the quality assessment of the source, and the further overall preliminary estimate of resources.

1.6 Mathematical Apparatus for Quantitative Interpretation

1.6.1 Redox Potential–Mathematical Expressions

1. General expression of transient response for in situ measurement of soil ORP:

$$U_r(t) = L - A_o e^{-\frac{t}{\tau}}$$

where A_o and τ are the linear functions (inversely proportional) of the soil reduced conductivity (C_r), resulting as indicative of metal mineralization very close to the surface (minimum values of A_o and τ, similar and repeated in adjacent stations), being the linear relationship between A_o and C_r (C_r to values between 25 and 105 mhos) as follows:

$$C_r = -2.9261A_o + 103.01$$

2. General expression of reduced Redox Potential dependence with depth (H), thickness (P), and metal concentration of the source:

$$U_r(H, P, C) = \left(212e^{-\frac{80}{H}}\right)H^{-0.15}e^{-0.0005H}e^{\frac{P}{H}}e^{C\frac{\Delta C}{\Delta H}} \qquad (1.1)$$

where $A = \left(212e^{-\frac{80}{H}}\right)$

The use of this general expression varies depending on the area or domain of application (oil, minerals, and environment/archeology), for which we consider three mean levels deep: 1000, 100, and 10 m, which when evaluated in the expression, results for the A coefficient values previously known, obtained empirically: 196, 95, and 0.07, respectively. The type of application (average depth of the source) also allows assuming considerations that simplify the expression. Thus, for oil, it can be assumed that the metal concentration into the source is constant ($\Delta C/\Delta x = \Delta C/\Delta y = \Delta C/\Delta z = 0$) and the thickness of the reservoir is very small as compared to its depth, and hence the expression to be read:

$$U_r = 196H^{-0.15}e^{-0.0005H},$$

which is valid for heavy hydrocarbons, where negative anomalies are characteristic of intensities greater than 30 mV. As for heavy-light and light hydrocarbons, where negative anomalies feature amplitudes smaller than 30 mV, the valid expression is

$$U_r = \left(5e^{-\frac{80}{H}}\right)H^{0.25}e^{0.0005H} \qquad (1.2)$$

For H = 1000 m the coefficient A is 4.6

In the case of applications for minerals and environmental/archeology, the above simplifications are no longer valid, so it is then necessary to have additional information on the thickness and concentration of the metal object in order to reduce ambiguity in interpretation.

1.6.2 Content of Chemical Elements—Mathematical Expressions

General expression of the normalized (Standardized) Content of chemical elements in soil dependence with depth, thickness, and metal concentration of the source:

$$C_n = \left(1039e^{-\frac{80}{H}}\right)\frac{1}{F}H^{-0.25}e^{-0.0005H}e^{\frac{P}{H}}e^{C\frac{\Delta C}{\Delta H}}, \qquad (1.3)$$

where F is the local background level.

The use of this expression as well as the one corresponding to ORP (1.1) depends on the field of application and, as it was in the previous case, the same three mean levels of depth were considered, 1000, 100, and 10 m, to evaluate A coefficient values: 958, 467, and 0.35, respectively. They are also the valid simplifications of the expression (1.3) for oil applications where, to be the unit the last two terms, it looks like

$$C_n = 958 \frac{1}{F} H^{-0.25} e^{-0.0005H}$$

Similarly, for mineral and environmental/archeological applications previous simplifications are invalid, resulting in a need for additional information to reduce ambiguity in interpretation.

1.6.3   Metal Source Quality

In oil applications an indicator parameter of hydrocarbon metal quality is defined. Its magnitude is being similar to the hydrocarbon API, and it is referred as Redox Grade or equivalent grade inorganic API:

$$^\circ REDOX = 2.69 F^{-0.25} e^{-0.0005F}, \tag{1.4}$$

where

$$F = F_{Ni} F_V F_{F_e} F_{Zn} \cdot 10^{-8} \tag{1.5}$$

The local background values of the various elements were being expressed in ppm.

In mineral and environmental/archeological applications as an indicator of quality of the surface metal source the expression currently used is

$$Ley_{med} = 10 C_{max}, \tag{1.6}$$

where C_{max} is the maximum value of the content in soil of the basic constituent element. In the case of Au, the multiplicative factor is 3.2, which is explained by its low solubility.

1.6.4   Relations Among Reduced or Standardized Attributes—Mathematical Expressions

It has been found that there is a linear dependence among the reduced spectral reflectance (RE_r) and the standard magnetic susceptibility (K_n) with the standard content of chemical elements in soil (C_n) resulting from the process of vertical migration of metal ions from the source. The dependence of RE_r shows a direct genetic link to the metal presence, which is not the case with K_n, whose link is indirect through the formation of the diagenetic magnetic minerals which originated as a result of the existence of the "Reducing Column". In the specific case of horizontal objectives (weathering crusts, oil reservoirs, pollution plumes at the ground water level, as well as horizontal asphaltic rocks) a linear dependence also occurs between the reduced Redox Potential (U_r) and the standard Chemical Content in soil. Such a linear attributive dependence is generally expressed in the form of up to three lines, characterizing the behavior of each attribute. Mathematical expressions for these linear relations are

$$C_n = \frac{b_i K_n - (b_i - a_i)}{a_i} \qquad (1.7)$$

$$C_n = \frac{b_i RE_r + a_i}{a_i} \qquad (1.8)$$

$$C_n = \frac{b_i U_r + a_i}{a_i}, \qquad (1.9)$$

being the latter valid only for horizontal targets, where

$$a_i = \frac{1}{C_n - 1}$$

and

$$b_i = \left\{ \begin{array}{c} \frac{1}{K_{ni}-1} \\ \frac{1}{RE_{ri}-1} \\ \frac{1}{U_{ri}-1} \end{array} \right\},$$

being i any point on the lines defining the linear dependence.

Considering the above relations, the more general expressions describing the dependence of K_n, RE_r, and U_r (horizontal objects) with parameters as lying, geometric, and metallic quality of the sources are the following:

$$K_n = \frac{b_i - a_i}{b_i} + \frac{a_i}{b_i}\left[\left(1039 e^{-\frac{80}{H}}\right)\frac{1}{F}H^{-0.15}e^{-0.0005H}e^{\frac{P}{H}}e^{\frac{C_{AC}}{AH}}\right] \qquad (1.10)$$

$$RE_r = \frac{a_i}{b_i}\left[\left(\left(1039 e^{-\frac{80}{H}}\right)\frac{1}{F}H^{-0.15}e^{-0.0005H}e^{\frac{P}{H}}e^{\frac{C_{AC}}{AH}}\right) - 1\right] \qquad (1.11)$$

$$U_r = \frac{a_i}{b_i}\left[\left(\left(1039 e^{-\frac{80}{H}}\right)\frac{1}{F}H^{-0.25}e^{-0.0005H}e^{\frac{P}{H}}e^{\frac{C_{AC}}{AH}}\right) - 1\right], \qquad (1.12)$$

being the latter valid only for horizontal objects.

From the above, expressions are derived corresponding for oil applications, from which the target depth is determined:

$$K_n = \frac{b_i - a_i}{b_i} + \frac{a_i}{b_i}\left(958\frac{1}{F}H^{-0.25}e^{-0.0005H}\right)$$

$$RE_r = \frac{a_i}{b_i}[(958\frac{1}{F}H^{-0.25}e^{-0.0005H}) - 1]$$

$$U_r = \frac{a_i}{b_i}[(958\frac{1}{F}H^{-0.25}e^{-0.0005H}) - 1]$$

For mineral applications, the expressions of horizontal, deep (100 m), and surface targets (10 m) are as below. Notice that the latter is also valid for environmental applications:

$$K_n = \frac{b_i - a_i}{b_i} + \frac{a_i}{b_i}\left(\left[\frac{467}{0.35}\right]\frac{1}{F}H^{-0.25}e^{-0.0005H}e^{\frac{p}{\bar{H}}}\right)$$

$$RE_r = \frac{a_i}{b_i}\left[\left(\left\{\frac{467}{0.35}\right\}\frac{1}{F}H^{-0.25}e^{-0.0005H}e^{\frac{p}{\bar{H}}}\right) - 1\right]$$

$$U_r = \frac{a_i}{b_i}\left[\left(\left\{\frac{467}{0.35}\right\}\frac{1}{F}H^{-0.25}e^{-0.0005H}e^{\frac{p}{\bar{H}}}\right) - 1\right]$$

These expressions allow the estimation of the thickness of the lens (its variability) under the conditions of a homogeneous metallic composition and a constant depth of the top of the source.

1.7   Empirical Foundation of the *Redox Complex*

Redox Complex has been applied in Cuba (Pardo 2001, 2003; Pardo et al. 1997; Pardo and Stout 1999, 2001; Pardo et al. 2000a, 2000b, 2001, 2003) for different control areas (with a great deal of knowledge on the objects of study) related to different types of oil fields (Cantel, Pina, Varadero-Varadero Sur, Jatibonico and Cristales) and metallic mineral deposits (Mella, La Union, Antonio, Little Golden Hill, Florencia-Cuerpos Norte y Sur, Jacinto-Vetas Beatriz y El Limón, Camagüey II, Cuba Libre-Río Negro, Yagrumaje Norte, Camarioca Este, Union 1, Descanso, Meloneras, Stock Guáimaro and Sigua), all of them the representatives of a wide variety of geological environments, genetic types, useful components, textural–structural features for both the reservoir and the ore, as well as geometric and lying (occurrence) conditions and topography. Applications in environmental and archeology consider, respectively, industrial perimeter zones such as Fábrica de Baterías Secas Pilas Yara, Siderúrgica Antillana de Acero, Refinería Ñico López,

Electroquímica de Sagua, Fábrica de Sulfometales, and Fábrica de Pinturas Capdevilla, as well as Presa de Colas de Mina Delita; and a few archeological sites among them are Jardín Exterior del Castillo de la Fuerza, Caimito, and Valle de los Ingenios. All of these investigations have served as an experimental basis for the development of the mathematical apparatus here presented.

1.8 Updating Issues

The unconventional and relatively new geochemical exploration techniques based on the *Principle of Vertical Migration of Metal Ions*, such as MMI (Canadian, 1990s) and CHIM (Russia, 1980s) are currently used as additional exploration tools to increase the effectiveness of geological prospecting. In this same exploration current *Redox Complex* is inserted.

1.9 Novelty

Redox Complex represents a cheaper alternative to the aforementioned geochemical techniques, since it does not require the use of selective extraction and ICP mass for analytical determinations, using, instead, conventional digestion and ICP spectral. Moreover, this complex is supported upon a mathematical apparatus for quantitative estimates of the metal sources and its effectiveness has been proven effective in other areas of application such as oil exploration (as MMI) as well as environmental and archeological studies.

1.10 Economic Assessment and Social Contribution

Redox Complex employee, with complementary character, within the minimum exploration/research complex, raises the effectiveness of geological investigations and reduces overall costs, allowing the optimization of decision-making process and improving the success rate of drilling and/or exploratory excavation.

1.11 Conclusions and Recommendations

Disclosed is a new unconventional geophysical–geochemical exploration tool for indirect detection, characterization, and evaluation of targets of metallic nature. Its resulted final product consists of a map depicting the vertical projection of the metal target; a central section with its geometrization, together with the behavior of the featured attributes; the parameters defining the nature and quality of metallic source; and a general resource estimation. It is recommended to extend the application of this exploration tool to other different areas, particularly those here proposed, to further validation.

References

Birell R (1996) MMI geochemistry: mapping the depths. Min Mag 174(5):306–307

Hamilton SM (1998a) Electrochemical mass-transport in overburden: a new model to account for the formation of selective leaches anomalies in glacial terrain. J Geochem Explor 63:155–172

Hamilton SM (1998b) New electrochemical studies by ontario geological survey, Explore No.101

Hamilton SM (2000) Spontaneous potentials and electrochemical cells. In: Geochemical remote sensing of the subsurface. In: Handbook of exploration geochemistry, vol 7. Elsevier, Amsterdam, pp 81–119

Hamilton SM, Cameron EM, Mc Clenaghan MB, Hall GEM (2004) Redox, pH and SP variation over mineralization in thick glacial overburden (I): methodologies and field investigation at the Marsh Zone gold property. Geochemistry 4(1):33–44

Kelley DL, Cameron EM, Southam G (2004) Secondary geochemical dispersion through transported overburden. SEG, Perth 4 pp

Mann AW (1997) The use of MMI in mineral exploration. In: 10th International gold symposium. Rio de Janeiro, Brazil

Mann AW, Birrell RD, Fedikov MHF, de Souza HAF (2005) Vertical ionic migration: mechanisms, soil anomalies and sampling depth for Mineral Exploration. In: Geochemistry: exploration, environment analysis, vol 5. Geological Society Publishing House, pp 201–210

Moon CJ, Khan A (1990) Mineral exploration. Min Annu Rev 1990:203

Pardo M (2001) Proyecto de Investigación-Desarrollo (227): Análisis teórico y diseño de los parámetros cuantitativos del Método Redox en áreas de control (inédito). IGP, La Habana

Pardo M (2003) Proyecto de Investigación-Desarrollo (244): Estimaciones cuantitativas por el Complejo Redox en su aplicación a la prospección geológica, los estudios medio–ambientales y arqueológicos (inédito). IGP, La Habana

Pardo M (2004) Informe Final del Proyecto I + D 244 "Estimaciones Cuantitativas por el Complejo Redox en su aplicación a la Prospección Geológica, los Estudios Medio-Ambientales y Arqueológicos", (inédito). Instituto de Geología y Paleontología, La Habana, 27 pp. (38)

Pardo M, Álvarez J, Echevarría G (1997) Técnicas geofísico-geoquímicas no convencionales para la prospección de hidrocarburos; progresos de su aplicación en Cuba. Memorias del III Congreso Cubano de Geología, vol. I. pp 547–550

Pardo M, Stout R (1999) El método de Potencial Redox en suelos y su aplicación combinada con la kappametría a los fines de la prospección geológica. Memorias del I Congreso Cubano de Geofísica, La Habana

Pardo M, Stout R (2001) El método de Potencial Redox en suelos y su aplicación combinada con la kappametría a los fines de la prospección petrolífera. Memorias del II Congreso Cubano de Geofísica, La Habana

Pardo M, Stout R, Aragón R, Álvarez E, Lugo R (2000a) El método de Potencial Redox en la prospección de minerales; dos ejemplos: el yacimiento Mella y el stockwork La Unión. Memorias del IV Congreso Cubano de Geología, La Habana

Pardo M, Stout R et al (2000b) Informe del TTP Aplicación de la medición in situ del Potencial Redox en suelos sobre objetivos de prospección de minerales metálicos e hidrocarburos (inédito). ONRM, La Habana

Pardo M, Stout R et al (2003) El *Complejo Redox*: Estado del Arte en el 2002. Memorias del V Congreso Cubano de Geología, La Habana

Pardo M, Stout R, Kessell E, Mugía M, Morales MA, Rojas F (2001) La nueva técnica geoquímica de Potencial Redox en suelos: Principios y Aplicaciones. Memorias del II Congreso Cubano de Geofísica, La Habana

Smee BW (2003) Theory behind the use of soil pH measurements as an inexpensive guide to buried mineralization, with examples. Explore No. 118

Chapter 2
Historical Development of *Redox Complex*

Abstract The *Redox Complex* (ORP, magnetic susceptibility, spectral reflectance, and soil geochemistry) is a complex of unconventional exploration techniques, used for indirect detection and evaluation of various metal targets, which is based on the *Geochemical Principle of Metal Ions Vertical Migration*. This complex is successfully applied in various fields: oil and gas and metal ores exploration; studies of oil and metal contaminants in soils; and the search for metallic archeological burials. The use of these techniques is intended to complement the conventional prospecting complex with the purposes of reducing areas and/or the selection of the most favorable targets, resulting in an increase in economical–geological effectiveness of investigations. The *Redox Complex* is implemented without physical or chemical affectation to the environment. The combined application of redox potential and magnetic susceptibility of soils to geological prospecting, which is its antecedent, is determined by the possibility of detecting the reducing column that is formed directly on metal and hydrocarbon occurrences and reaches the upper part of the section. Within this column, the conversion of nonmagnetic iron minerals to more stable magnetic varieties is favored, which explains the observed inverse correlation between the attributes and justifies integration of methods. The results of research in control areas confirm the fact, previously noted, that ORP in soil is insensitive to variations in topography and tectonics, being the potential background level a function, only, of the type of geological environment. The only noisy sources known to date are the dry marshy areas or other accumulations of organic material and alluvium with high content in metals, which are reducing surface geochemical targets that produce characteristic intense anomalies (<-100 mV) of steep and symmetrical gradients, occasionally identifiable from direct observation in the field. Although available applications are still, statistically, insufficient for experimental foundation of empirical regularities mathematically formulated, however, they allow, in principle, the establishment of a set of possibilities and limitations of *Redox Complex* for each area of application.

Keywords Geological prospecting · Soil redox potential · Magnetic susceptibility · Reduced chimneys · Hydrocarbon deposits · Metallic mineral occurrences

© The Author(s) 2016

M.E. Pardo Echarte and O. Rodríguez Morán, *Unconventional Methods for Oil & Gas Exploration in Cuba*, SpringerBriefs in Earth System Sciences, DOI 10.1007/978-3-319-28017-2_2

19

2.1 Introduction

The historical—practical evolution of redox potential (ORP) in soils, used in combination with kappametry for the purpose of hydrocarbon and metal ores exploration (Pardo y Stout 1999), since its innovation and introduction in 1992 (Alfonso et al. 1993) until 2001 (Pardo y Carballo 1996; Pardo y Domínguez 1997; Pardo et al. 1997, 2000, 2001a, b; Pardo y Stout 1999, 2001), required a new theoretical and experimental stadium led to multiparametric complementation, the development of the theoretical basis of the technical complex, and the implementation of its quantitative interpretation. Empirical support was needed for that, with minimal (still insufficient) volume of observations on known targets or control areas, which would be conducted in two research projects for 2001–2002 (Pardo 2002) and 2003–2004 (Pardo 2004). The areas were selected to characterize response to different types of environments and deposits with different characteristics: of lying (occurrence), geometry, and composition; so it could do more representative analysis for design quantitative interpretation parameters.

2.2 Historical Development of *Redox Complex*

Progress in the historical development of *Redox Complex* has followed the path from empirical observation, through experimentation and application of mathematical expressions that model the observed regularities, to an approach to the theory. The main stages of this development are:

1991–1993

- Invention, design, and introduction of in situ soil Redox Potential measurement.
- Innovation, design, and introduction of airborne gamma-spectrometric (AGE) scenarios for oil: The integrated index K/eTh.

1996–2001

- Establishment of the genetic relationship between ORP reduced by local background and magnetic susceptibility of the soil from the *Principle of Metal Ions Vertical Migration*. Introduction of Uredox–Kappa complex to geological prospecting.
- Invention, design, and introduction of ORP temporal measurement from the mathematical formulation of the transient behavior.
- Establishment of the genetic relationship between ORP reduced by local background and Soil Geochemistry from the *Principle of Metal Ions Vertical Migration*. Introduction of Uredox–Kappa–Soil Geochemistry complex to geological prospecting.

2002–2005

- Establishment of the genetic relationship between spectral reflectance reduced by local background in soil samples and soil geochemistry, from the *Principle of Metal Ions Vertical Migration*. Innovation, design, and introduction of satellite scenarios for oil and metallic minerals.
- Introduction of normalization (standardization) by local background of the magnetic susceptibility and soil geochemistry.
- Final *Redox Complex* integration: Satellite scenarios, redox potential, magnetic susceptibility, spectral reflectance, and soil geochemistry (attributes reduced or normalized by local background).
- Invention, development (modeling), and automated deployment of mathematical expressions that link various attributes with lying, geometry, and composition parameters of prospecting targets.
- Innovation, design, and introduction of *Redox System* for geological applications (oil and minerals), environmental, and archeological studies.

2010–2013

- An approach to the theory of the processes controlling metal mobilization, transport, and accumulation in surficial environment on buried ore bodies and hydrocarbon deposits.

Considering its beginnings, applications to geological prospecting of ORP in soils method are determined by the possibility of detecting the reducing environment column that is formed directly on metal and hydrocarbon occurrences and reaching the top of the section. Within this column, the conversion of nonmagnetic iron minerals to more stable magnetic varieties is favored, which explains the observed inverse correlation with magnetic susceptibility and justifies the integration of both methods. The use of these techniques is intended to complement the conventional prospecting complex with purposes of reducing areas and/or selection of the most favorable targets, resulting in an increase in economical–geological effectiveness of investigations.

Redox Potential (ORP) in soils in situ measurement was a little-known method in the 1990s, determination that was conventionally performed on aqueous or semi-aqueous media (water, sludge, and soil solutions). Its application alone or in combination with soil kappametry for geological exploration had not been since reported. The beginnings of the development and introduction of this method in Cuba were dating from 1992 to 1993 (Alfonso et al. 1993), where it was used in combination with other nonconventional techniques to reveal surface areas of reducing environment on hydrocarbon occurrences. Subsequently, it has been applied in combination with soil kappametry on various metallic and hydrocarbon targets (Pardo y Carballo 1996; Pardo et al. 1996; Pardo y Domínguez 1997; Pardo et al. 1997, 2000, 2001a, b; Pardo y Stout 1999, 2001; Pardo et al. 2003). For its part, the soil kappametry with purposes of geological prospecting has its antecedents in the detection of dispersion haloes on metal occurrences, although it was also known in oil exploration (Saunders et al. 1991). This technique was directed to the detection of anomalies in the magnetic susceptibility in a given soil horizon, from measurements in samples for other method (soil geochemistry samples or samples for analysis of hydrocarbon gases).

The geological assumptions underlying the combined application of soil ORP and magnetic susceptibility (Kappametry) methods for geological exploration are:

- Prospecting for oil and gas.

  It is recognized (Saunders et al. 1991) the fact of the vertical migration of light hydrocarbons (microseepages as ultra-small gas bubbles) from the hydrocarbon deposit in depth, through fractures, layer planes, and faults to the soil top level. For long periods of time, the alterations related to microseepages may take place in the upper levels of the soil from a range of reactions either in gaseous or aqueous phase. Microbial action on the light hydrocarbons was produced as by-products carbon dioxide and hydrogen sulfide, determining a column of chemically reducing environment on the occurrence. Moreover, this environment favors the conversion of nonmagnetic iron minerals (hematite) in magnetic oxide (magnetite) and magnetic sulfides (pyrrhotite and griegite), a fact that underlies the correlation of ORP anomalies (lows) with those of the magnetic susceptibility (highs). On the other hand, concentrations of carbon dioxide in groundwater form carbonic acid which can react with the clay minerals to create secondary mineralization of calcium carbonate and silicification, resulting in more dense and resistant to erosion surface materials (leading to geomorphic anomalies). According to Saunders et al. (1993), simultaneously, the destruction of clay minerals by carbonic acid and organic acids can release potassium, which is subsequently leached. Conversely, thorium remains relatively fixed in its original distribution within the insoluble heavy minerals; hence, it observed characteristic anomalies (lows) of the relationship K/eTh on hydrocarbon deposits.

- Metallic mineral prospecting.

  It is the known (Hernán Vázquez 1997) process of vertical migration of metal ions from an occurrence in depth to the surface, determining the formation of a reducing environment column thereon. The metal transport involves diffusion or hydromorphic transport mechanisms linked to weathering events, among others, a process that also affects newly transported cover (exotic). As in the previous case, the reducing environment promotes the conversion of nonmagnetic iron minerals into more stable magnetic varieties.

2.3 Methodological Aspects

According to Pardo y Stout (1999), for the soil ORP in situ measurement two electrodes connected to a digital millivolt meter of high input impedance (sensitivity 0.1 mV; commercial) are used: one inert platinum, and other reference copper (nonpolarizable electrode; commercial), located immediately next into a hole of 10–30 cm deep. The ionic communication which closes the circuit is ensured through the porous ceramic of the reference electrode (Fig. 2.1).

Fig. 2.1 Measurement of Redox Potential in soils

The measurement with the disclosed device has a transient behavior, determining the potential by an algorithm from five readings with a time difference between them of a minute. The quality of field observations is assessed using the absolute error in the determination of the potential, considering 10 % of control measurements made in the same holes used for routine measurements. The acceptable accuracy for the described applications should not exceed 15 mV.

The Magnetic Susceptibility measurement is made with a Kappameter KT-5 (sensitivity $1 \cdot 10^{-5}$ SI), making seven readings distributed over the floor and walls of the hole, which are averaged. The quality of the field observations is assessed from the relative error in determining the mean value of kappa, considering 10 % of control measurements in the same holes. The accepted accuracy must not exceed 15 % (Fig. 2.2).

Thus, the method of Redox Potential in soils and its combined application with kappametry for the purposes of geological prospecting offered a priori some advantages:

- The possibility of indirect detection of deep or conceal metal occurrences and hydrocarbon reservoirs (structural and stratigraphic traps) with high planimetric accuracy on their location.
- Unlike spontaneous potential method, the measurements are not affected by other processes such as electrofiltration, electrodiffusion, electrokinetic coupling, and bioelectric activity of vegetation.
- Integration with the measurement of soil magnetic susceptibility increases the informational value of the attribute and hence its geological effectiveness.

Fig. 2.2 Measurement of the Magnetic Susceptibility in soils

- As a complement to other conventional geochemical techniques (e.g., soil geochemistry), it raises its spatial resolution, thereby increasing the success rate of exploratory drilling.
- Measurements are not apparently affected by surface transport processes.
- The simplicity and low cost of field operations.

2.4 Research Objects

Research in control areas, in a first project (Pardo 2002), was developed on four objects of hydrocarbons, six of metallic minerals, three of environmental studies, and two of archeology. In a second project (Pardo 2004), investigations were carried on two objects of hydrocarbons, one of metallic minerals, three of the environment studies, and one of the archeologies. In all cases, the technical complex used (*Redox Complex*) considered redox potential, magnetic susceptibility, spectral reflectance, and soil geochemistry. Mineralogical study was only occasionally used in cases where necessary. Chemical analyzes by ICP spectroscopy considered useful and accompanying metallic elements of each occurrence, metal elements characterizing hydrocarbons, the main pollutants of the studied industries, and a typical spectrum of metallic elements for archeology. Thus, the *Redox*

Complex was successfully used in different types of hydrocarbon reservoirs (Varadero—Varadero Sur, Cantel, Pina, Jatibonico, and Cristales Oil fields) and metallic minerals (Antonio, Little Golden Hill, Florencia—Cuerpos Norte y Sur, Jacinto—Vetas Beatriz and El Limón, Camagüey II, Cuba Libre-Río Negro, and Camarioca Este ore deposits), which are representatives of a variety of geological environments, structural types, genetic types, composition, structure–textural features of the reservoir and/or ore, lying (occurrence) conditions, and relief. Applications for environmental and archeological studies, respectively, considered the perimeter zones of industries: Fábrica de Baterías Secas Pilas Yara, Siderúrgica Antillana de Acero, Refinería Ñico López, Distribuidora de Hidrocarburos Güines, Distribuidora de Hidrocarburos Artemisa, Fábrica de Sulfometales, Fábrica de Pinturas Capdevilla, and Presa de Colas Delita; and the archeological sites: Jardín Exterior del Castillo de la Fuerza, Caimito, and Ingenio Guáimaro.

For the purposes of hydrocarbons (heavy and light, with depths between 300 and 2000 m) works were developed by isolated profiles, step 200 m, extending between 2 and 5 km. For purposes of metallic minerals (Kuroko VMS, Epithermal low and high sulfidation, Refractory chromite, Cyprus VMS, and Nickeliferous crust) works were developed by isolated profiles, step 20 m, extending from 200 to 600 m. For the purposes of environmental studies (surficial and occasionally at groundwater table contaminations of Fe, Mn, As, Pb, Zn, and hydrocarbons) works were developed by isolated profiles, step 20 m, with approximate length of 200 m. For the purposes of archeology (Fe and Au (As) metallic burials) works were developed by isolated profiles, step 0.5 m, with approximate length of 10 m.

2.5 *Redox Complex* Economic–Technical Data

Current estimates of productivity and cost of *Redox Complex* (in situ measurements of soil ORP and magnetic susceptibility and soil sampling) applications in mineral and environmental for terrains with difficulty category II–IV, observation step 20–25 m, respectively, are 25–17 PP/DC (500–340 m) and $250.00–480.00/km. Estimates of spectral reflectance, respectively, are 120 samples/DC and $3.50/sample ($175.00/km). The chemical analysis (samples analyzed in Central Laboratory of Mineral, LACEMI) has a response speed exceeding 30 days, with an average cost of $70.00–90.00/sample. Similarly, for hydrocarbon exploration, estimates of *Redox Complex* in vehicle-transported itineraries, step 200–250 m, are 20 PP/DC and $250.00/km, while profiling (pedestrian), 200–250 m step, difficulty category II, are 15 PP/DC and $400.00/km. Estimates of spectral reflectance and chemical analysis remain the same. All the above estimates do not consider the costs of mobilization–demobilization campaign, food, accommodation, and transportation of executioner staff.

2.6   Results

The results of research in control areas (Pardo 2002, 2004) confirm the fact, noted above, that the technique of soil ORP is insensitive to variations in topography and tectonics, being the potential background level a function, only, of the type of geological environment. The only sources of noise known to present are dry swampy wetlands or other accumulations of organic material and alluvium high in metals, which are reducing surface geochemical objectives that produce characteristic intense anomalies (<-100 mV) of steep and symmetrical gradients, occasionally identifiable from direct observation in the field.

Although the available applications (Pardo 2002, 2004) are still, statistically, insufficient as experimental foundation of empirical regularities mathematically formulated, however, they allow, a priori, the establishment of a set of possibilities and limitations of *Redox Complex* for each area of application:

• **Hydrocarbon Exploration**

(a) Possibilities:

  • Recognition and detailed mapping of the occurrence vertical projection and assessment, in an indirect and approximate way, of the oil quality.
  • Approximate determination of the depth of the occurrence and the possible structural trap type.
  • Approximate volumetric assessment of the occurrence from the detailed mapping results.

(b) Limitations:

  • Existence of several overlapping hydrocarbons levels (includes gas caps).
  • Failure to seal or it is much fractured.
  • Presence of surface reducing zones of various kinds.
  • Prevalence of the lithological response of magnetic susceptibility and spectral reflectance.
  • Existence of a recent transported cover (less than 10 years).

• **Metallic Mineral Exploration**

(a) Possibilities:

  • Recognition and detailed mapping of the occurrence vertical projection and rough assessment of its composition and quality.
  • Approximate determination of the depth, dip, and vertical extent of the main ore body.
  • Rough assessment of resources.

(b) Limitations:

- Existence of overlapping primary ore bodies.
- Presence of surface reducing zones of various kinds.
- Prevalence of the lithological response of magnetic susceptibility and spectral reflectance.
- Existence of a recent transported cover (less than 10 years).

- **Environmental Studies**

(a) Possibilities:

- Recognition and detailed cartography of the contaminated area, allowing to establish its nature (surface or subsurface) and grade.
- Approximate determination of the depth of the subsurface contaminant source.

(b) Limitations:

- Presence of surface reducing or oxidizing zones of various kinds.
- Prevalence of the lithological response of magnetic susceptibility and spectral reflectance.
- Existence of a recent transported cover (less than 10 years).

- **Archeological Studies**

(a) Possibilities:

- Planimetric position of metal burials and assessment of its composition.
- Approximate determination of the depth of objects.

(b) Limitations:

- Presence of surface reducing or oxidizing zones of various kinds.
- Prevalence of the lithological response of magnetic susceptibility and spectral reflectance.
- Existence of a recent transported cover (less than 10 years).

Below are presented, in graphical form, the most representative results of the application of *Redox Complex* in some studied control areas and it is commented on the regularity in the spatial behavior of the different measured attributes.

Oil and Gas

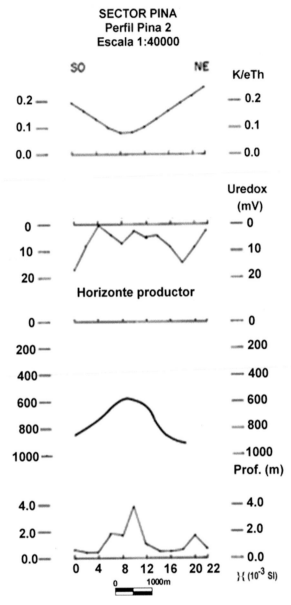

SECTOR PINA
Perfil Pina 2
Escala 1:40000

At the *Pina* light oil field, spatial correspondence of a minimum of K/eTh relationship with a maximum of magnetic susceptibility and a disjointed decreased field values in the reduced ORP, linked to the apical part of the producer structure, is observed. Spectral reflectance was not determined.

Metallic Minerals

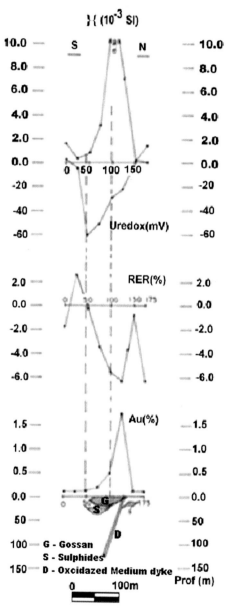

SECTOR GOLDEN HILL
Perfil Little Golden Hill
Escala 1:5000

At the *Little Golden Hill* epithermal high-sulfidation gold deposit, spatial correspondence of a maximum of magnetic susceptibility and minimum values of

reduced spectral reflectance (darkening of the ground) to the oxidation zone of the
deposit is observed. The reduced ORP low corresponds to the massive sulfide area
in the depth (gossan below).

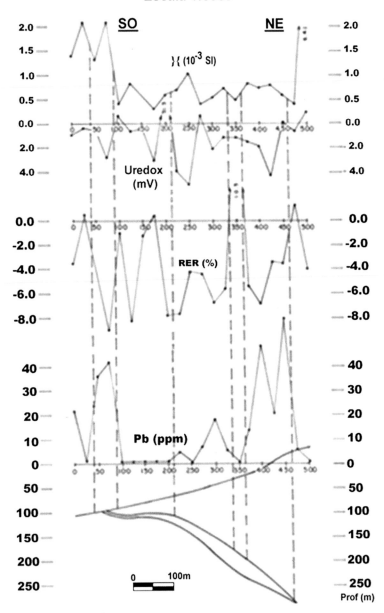

At the *Antonio* polymetallic Kuroko-type volcanogenic massive sulfide deposit, spatial correspondence of magnetic susceptibility highs with lows of reduced ORP and lows of reduced spectral reflectance, associated with increments of Pb in soil (from the deep), is observed.

Environmental studies

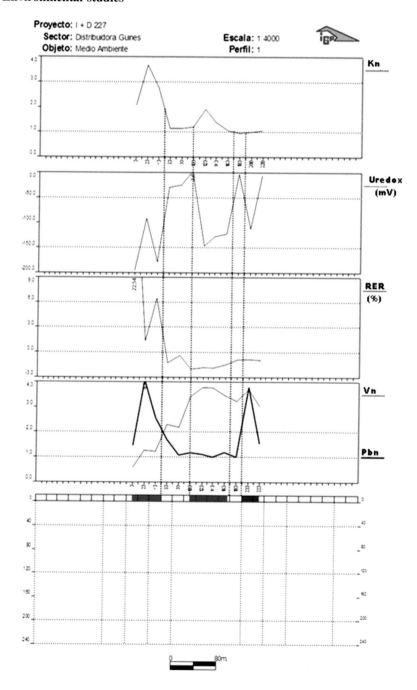

At the *Distribuidora de Hidrocarburos Güines*, not observable surface contamination, and presumably contamination at the level of the water table (from
100 m in the profile), is expressed by the spatial correspondence of maxima of the
standard magnetic susceptibility, minimum values of reduced ORP, maxima and
minima of the reduced spectral reflectance, and a corresponding increment (normalized) of the metallic elements hydrocarbon indicators: light (Pb) and heavy (V).

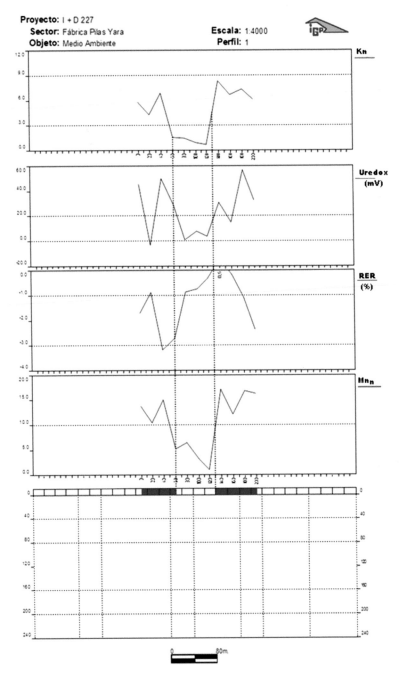

At the *Fábrica de Baterías Secas Pilas Yara*, unobservable manganese and other surface metal contamination is expressed by the spatial correspondence of maxima normalized magnetic susceptibility, maximum values of reduced ORP (oxidation zones), an undifferentiated character of reduced spectral reflectance, and a corresponding (normalized) increment of the main contaminant (Mn).

Archeology studies

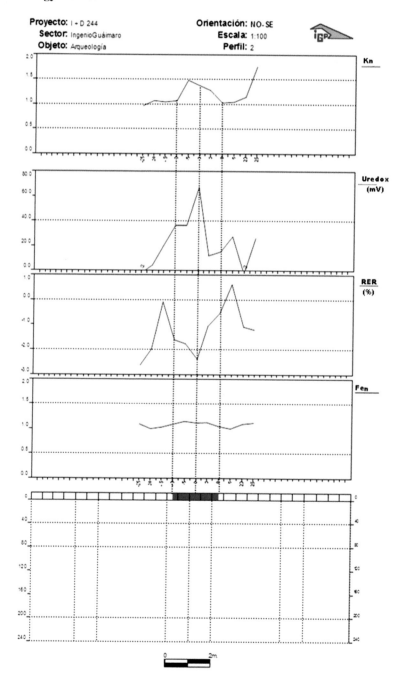

Proyecto: I - D 244   Orientación: NO-SE
Sector: IngenioGuáimaro   Escala: 1:100
Objeto: Arqueología   Perfil: 2

Kn

Uredox (mV)

RER (%)

Fen

0   2m

At the *Ingenio Guáimaro* (Valle de los Ingenios, Sancti Spiritus), burial of a Jamaican Train is recognized by the spatial correspondence of a maxima normalized magnetic susceptibility, a maxima of reduced ORP (oxidation zone), a minima of reduced spectral reflectance, and the corresponding (normalized) increment of indicator element (Fe).

2.7 Conclusions

The combined application of redox potential and magnetic susceptibility of soils to geological prospecting, which is the antecedent of *Redox Complex*, is determined by the possibility of detecting the reducing environment column that is formed directly on metal and hydrocarbon occurrences and reaches the top of the section. Within this column, the conversion of nonmagnetic iron minerals to more stable magnetic varieties is favored, which explains the observed inverse correlation between the attributes and justifies integration of methods.

The multiparameter complementation and development of a theoretical basis of the technical complex, based on empirical support (still insufficient) of observations on known targets or control areas, identify the *Redox Complex* (ORP, magnetic susceptibility, spectral reflectance, and soil geochemistry) as a unconventional exploration technical complex, used for the indirect detection and evaluation of various metallic objects. It is based on the *Geochemical Principle of Metal Ions Vertical Migration* and successfully applied in various fields: oil and gas and metallic minerals exploration; studies of oil and metal contaminants in soils; and the search for metallic archeological burials. The use of these techniques is intended to complement the conventional prospecting complex with purposes of reducing areas and/or selection of the most favorable targets, resulting in an increase in economical–geological effectiveness of investigations.

References

Alfonso JR, Pardo M y otros (1993) Informe sobre los trabajos metodológicos–experimentales de métodos geofísicos y geoquímicos no convencionales para la prospección de hidrocarburos someros en Cuba Septentrional (inédito). ENG, La Habana

Hernán Vázquez C (1997) Exploración Geoquímica. Revista Eafit 107.2

Pardo M (2002) Informe del Proyecto I + D 227: Análisis teórico y diseño de los parámetros cuantitativos del método Redox en áreas de control (inédito). IGP, La Habana

Pardo M (2004) Informe del Proyecto I + D 244. Estimaciones cuantitativas por el Complejo Redox en su aplicación a la prospección geológica, los estudios medio–ambientales y arqueológicos (inédito). IGP, La Habana

Pardo M, Carballo O (1996) Mediciones experimentales in situ del Potencial Redox en suelos sobre objetivos meníferos en los sectores de Loma de Hierro y Loma Roja, Provincia de Pinar del Río (inédito). IGP, La Habana

Pardo M, Domínguez E (1997) Reporte sobre los trabajos de Potencial Redox, Susceptibilidad Magnética y Análisis Químicos de suelos, para petróleo, en un área del Sur de Camagüey (Cuenca Vertientes) (inédito). IGP, La Habana

Pardo M, Rojas F y otros (1996) Naturaleza geoquímico-mineralógica de las anomalías aerogammaespectrométricas vinculadas con depósitos de hidrocarburos; primeros resultados en Cuba: Pina- Ciego de Ávila (inédito). IGP, La Habana

Pardo M, Stout R y otros (2000) Informe del TTP Aplicación de la medición in situ del Potencial Redox en suelos sobre objetivos de prospección de minerales metálicos e hidrocarburos (inédito). ONRM, La Habana

Pardo M, Stout R y otros (2003) El Complejo Redox: Estado del Arte en el 2002. Memorias del V Congreso Cubano de Geología, La Habana

Pardo M, Stout R (1999) El método de Potencial Redox en suelos y su aplicación combinada con la Kappametría a los fines de la prospección geológica. I Congreso Cubano de Geofísica, La Habana

Pardo M, Stout R (2001) El método de Potencial Redox en suelos y su aplicación combinada con la Kappametría a los fines de la prospección petrolífera. II Congreso Cubano de Geofísica, La Habana

Pardo M, Álvarez J, Echevarría G (1997) Técnicas geofísico-geoquímicas no convencionales para la prospección de hidrocarburos; progresos de su aplicación en Cuba. III Congreso Cubano de Geología, La Habana

Pardo M, Stout R, Aragón R, Álvarez E, Lugo R (2001a) El método de Potencial Redox en la prospección de minerales; dos ejemplos: el yacimiento Mella y el stockwork La Unión. IV Congreso Cubano de Geología, La Habana

Pardo M, Stout R, Kessell E, Mugía M, Morales MA, Rojas F (2001b) La nueva técnica geoquímica de Potencial Redox en suelos: Principios y Aplicaciones. II Congreso Cubano de Geofísica, La Habana

Saunders DF, Burson KR, Thompson CK (1991) Relationship of soil magnetic susceptibility and soil gas hydrocarbon measurements to subsurface petroleum accumulations. Bull Am Ass Petr Geol 75:389–408

Saunders DF, Burson KR and others (1993) Relation of thorium-normalized surface and aerial radiometric data to subsurface petroleum accumulations. Geophysics 58(10):1417–1427

Chapter 3
Processes Controlling Metal Mobilization, Transport, and Accumulation in the Surficial Environment over Buried Ore Bodies and Hydrocarbon Deposits: A Review

Abstract Considering the relevance of the *Geochemical Principle of Mobile Metal Ions Vertical Migration* (MMIVM), in which many up-date geochemical and geophysics–geochemical exploration techniques (i.e., electrochemical CHIM, MMI™, SGH, *Redox Complex*) are based, a summary review on different theoretical aspects related to metal mobilization, vertical transport to the surface, and the resultant accumulation of these vertically transported metals in surficial media from buried ore bodies and hydrocarbon deposits is made. As conclusion, it is outlined the general features characterizing these processes: microbial activity and water–rock reactions with gas (hydrocarbons, N, CO_2, H_2, and others) generation during target oxidation; ascending reduced gas microbubbles (colloidal size) with reduced metal ions attached, which results in 'reduced chimneys' reaching the surface; barometric pumping and capillary rise moving upward ions and submicron metal particulates, into the unsaturated zone; redistribution of ions in the near-surface environment by downward-percolating groundwater (after rainfall), as well as by the upward effects of evaporation and capillary rise, all of which explain the soil metal accumulations in a very shallow (10–30 cm) 'metal accretion zone'.

Keywords Mobile metal ions (MMI) · Geochemical exploration · Microbial activity · Reduced chimneys · Buried ore bodies · Hydrocarbon deposits

3.1 Introduction

Empirical observations, particularly that using high-resolution soil geochemistry as an analytical tool, suggest that soil geochemical anomalies over mineralization and hydrocarbon deposits are dynamic, and maintained by an ascending supply of ions from the source. This has been particularly evident in areas of exotic overburden. Research to define a unique and coherent theory for the mobilization and vertical migration of ions to form geochemical (and some geophysical) anomalies over concealed ore bodies, and hydrocarbon deposits, is still an updating task. It is

© The Author(s) 2016   37
M.E. Pardo Echarte and O. Rodríguez Morán, *Unconventional Methods for Oil & Gas Exploration in Cuba*, SpringerBriefs in Earth System Sciences, DOI 10.1007/978-3-319-28017-2_3

thought that no single mechanism shown to be capable of adequately explaining soil geochemical anomaly development in all geological situations, given the wide variety of terrains and climates that occur and the different conditions encountered beneath and above the water table.

In the past, there has been a concerted effort toward examining the processes of metal mobilization, the mechanisms of metal transport (including metals associated to hydrocarbon microseepage) through the overburden cover, and the resultant accumulation of these vertically transported metals in surficial media. In consequence, many well-documented related facts could be stated, although there is still debate regarding the factors controlling them, for instance:

- The presence of secondary geochemical anomalies in transported overburden above buried mineral bodies.
- Gases (hydrocarbons and high concentrations of N, CO_2, H_2, and others) over buried ore bodies resulting from the long-term water–rock reactions among ore bodies and their host rocks, and also from bacteria activity during ore oxidation.
- Research and case studies over known metal/hydrocarbon-bearing zones showing that mobile metal ions and additional compounds accumulate in surface soils above these "metal sources."
- The process of vertical migration of chemically reduced gases and metals from the oxidizing buried feature (ore bodies and hydrocarbon deposits) to surface, which results in 'reducing chimneys' reaching the surface, 'carbonate and acidic caps' in the unsaturated zone, and metals accumulating in a very shallow 'metal accretion zone' in soils.

Considering the relevance of the **Geochemical Principle of Mobile Metal Ions Vertical Migration** (MMIVM), in which many up-date geochemical and geophysics–geochemical exploration techniques (i.e., electrochemical **CHIM, MMI, SGH, Redox Complex**) are based on a summary review of different aspects on the theories related to metal mobilization, vertical transport to the surface, and the resultant accumulation of these vertically transported metals in surficial media from buried ore bodies and hydrocarbon deposits is presented.

3.2 Summary on Some Geochemical and Geophysics–Geochemical Exploration Techniques Used for the Study of Concealed Ore Bodies and Hydrocarbon Deposits

According to Fabris and Keeling (2006), the electrochemical technique **CHIM** (Chastichnoe Izvlechennye Metallov), initially developed in Russia, is based on the migration of fast ions from an ore body and transported to the surface by a vapor/gas (Goldberg 1998). In this technique, a direct current applied on the ground allows to collect the ions in specially designed electrodes, from a much larger volume than would be possible with traditional methods of soil geochemical sampling.

In the technique of mobile metal ions (**MMI**), which uses partial leaching (Birell 1996), the charged particles arrive into surface soils (where they occur in very low concentrations) from ore bodies and hydrocarbon accumulations, using one or several transport mechanisms proposed by various authors (diffusion, vapor/gas, electrochemical, capillary rise, convection, and barometric or seismic pumping).

In the technique soil gas hydrocarbons (**SGH**) (Sutherland 2011), heavier hydrocarbons (in the C_5–C_{17} carbon series range) released by microbial action in the deep ore bodies and hydrocarbon accumulations migrate to the surface where they are sampled and measured in a gas chromatograph/mass spectrometer. According to the author, dissimilar combinations and specific proportions of organic hydrocarbons define different signatures directly related to the objective under study, allowing not only the location but also the identification of the type of target that may be present.

Hamilton (2007) recommend the simultaneous use of three classes of geochemical tools in order to optimize the effectiveness of geochemistry in discrimination and prioritization of buried targets: selective leaches (MMI^{TM} and Enzyme Leach$^{®}$), soil hydrocarbon techniques (SGH and Soil Desorption Pyrolysis (SDP)), and pH slurry analysis. According to this author, MMI^{TM} and Enzyme Leach$^{®}$ measure the primary elemental signal given off by buried mineral deposits, but they can also detect a secondary signal caused by major geochemical processes occurring in overburden above the deposit, and related to the buried feature. Among these secondary processes an important role is played by the acidification due to metal oxidation above the water table. Besides, soil hydrocarbon techniques are useful to measure increases in heavier hydrocarbons and other gases in soils above mineral deposits. The changes are interpreted to be related to increases in autotrophic microbes that thrive in a 'reducing chimney' occurring in overburden above the mineral deposit. Finally, soil slurry pH seeks to detect pH responses that are known to occur over mineral deposits. Results can take the form of an acidified area above the water table (the 'acidic cap') or an alkaline area beneath it.

According to Pardo et al. (2003); Pardo and Rodríguez (2009), the **Redox Complex** (Redox Potential, Magnetic Susceptibility, spectral reflectance and soil geochemistry/total leach) is a complex of nonconventional techniques of exploration used (with complementary character within the conventional exploration complex) for indirect detection and evaluation of various targets of metal nature. It is based on the **Geochemical Principle of Mobile Metal Ions Vertical Migration** (MMIVM) (ions released from metal targets at depth), offering information on the terrain modifications taking place in its uppermost part (10–30 cm of the soil profile) over the target. This complex has been applied, with success, in different spheres, e.g., exploration of hydrocarbons and metal minerals; studies of metal and hydrocarbon contamination in soils and subsurface; and the search of metal archeological burials. The **Redox Complex** improves the likelihood of success since each method detects a different physical or chemical aspect resulting from a single, large-scale microbial/geochemical process occurring over buried mineralization and hydrocarbon deposits: 'reducing chimneys' reaching the surface; change of soil magnetic susceptibility and spectral reflectance; and metals accumulating in a very shallow (10–30 cm) 'metal accretion zone' in soils. Notwithstanding its potential, there are many possible sources of error when using the *Redox Complex*: physical attributes, as soil magnetic susceptibility and spectral reflectance, are somehow lithological dependent; redox potential is dependent for surficial reducing or oxidized zones of different natures; and soil geochemical signal from mineralization or hydrocarbon target is usually subtle, and some surficial processes can produce equivalent or stronger responses. (According to Hamilton 2007, one of the most likely sources of error of soil geochemistry in drier terrain is variable moisture content due to drier soils which are associated to more acidic pH and higher background metal concentrations—both in humus and mineral soils). The *Redox Complex* is appropriate for target discrimination and prioritization of numerous geophysical targets. However, it is less suitable for target generation, i.e., the observation of large grids, because of its low productivity (a short sample spacing is used, no more than a half of the anticipated width of the target, and the measure of redox potential is time consuming), and also owing to the many 'false anomalies' related to variable surficial conditions. In general, conscientious observers, good field notes, and—during interpretation stage—a basic knowledge of surficial geophysics–geochemical processes are critical success factors.

3.3 Summary on Some Theories Explaining Surface Anomalies Over Deeply Buried Mineralization

Several researchers have proposed various transport mechanisms to explain the ionic migration to the surface from the deeply buried mineralization, among which are diffusion, vapor/gas, electrochemical, capillary rise, convection, and advection water/gas (seismic and barometric pumping).

Diffusion, generally, is discarded by most researchers to be too slow, although Smee (1998) noted that hydrogen ions can diffuse fast enough to be involved in the chemical redistribution that occurs in alteration halos above mineral and hydrocarbon deposits.

In the vapor/gas transport ions are attached, from a hydrophilic view, to the gas bubbles below the water table and dragged by these (Kristiansson and Malmqvist 1984; Mann et al. 1995). For his part, Goldberg (1998) has suggested the presence of a kind of 'fast ions' moved upward by gases.

Various electrochemical models have been proposed (Sato and Mooney 1960; Govett 1973; Hamilton 1998, 2000), existing little doubt about the electrochemical effects in either part or all ore buried occurrences. However, the fact that if they are a universal cause or a result of ion migration is still under discussion. In general, all models show an electrochemical cathode zone at the top of the buried conductor. The model developed by Hamilton (1998) proposes the propagation upward of reduced species form a reduced column on the mineralized zone and develop geochemical anomalies in the overlying surface coverage. The theoretical rates of migration of ions in electrochemical fields are much faster than diffusion rates, and they are compatible with the formation of surface geochemical anomalies in young (~ 8.000 years) thicker layers of glacial sediments covering mineral deposits (Hamilton 1998; Hall et al. 2004).

The capillary rise, proposed by Mann et al. 1997, is considered an important process in the distribution and redistribution of ions in the environment close to the surface (above the water table). Empirical observations suggest that anomalies are preferably located 10–25 cm below the soil interface. In fact, experimental laboratory work developed by Mann et al. 2005 leads the authors to consider that perhaps the percolation of groundwater after rain, as well as the effects of capillary rise may explain many features of ion site in soils.

For arid or semiarid terrains, with a thick unsaturated zone, Cameron et al. (2004) intended as most likely transport mechanism of ions from the source, the mass transfer of groundwater or air, along with their dissolved or gaseous components (advective transport). Examples of this type of transport are mineralized groundwater pumping to the surface during seismic activity (seismic pumping), and the extraction of air plus gas by the barometric pumping. This type of mechanism requires the neotectonic fracture of coverage.

Also, the work of laboratory modeling of Mann et al. (2005) allowed suggesting the convection from the heat produced by the oxidation of the deposit as a possible mechanism of transport of ions below the water table.

Klusman (2009) proposes a model with two different transport mechanisms: for the saturated zone, a reducing chimney above ore deposits, emphasizing the role of reduced hydrocarbon gases, by microbial oxidation, in the formation of the chimney; and for the unsaturated zone, the vertical transport of ultrafine metal particles, applying the barometric pumping, which causes pressure gradients capable of moving submicron particles of Au to the surface.

3.4 Summary on Theories Explaining Surface Anomalies Over Hydrocarbon Deposits

Hydrocarbon microseepage is conceived as vertical hydrocarbon migration (mainly the series C_1-C_5, according to Price 1985) from hydrocarbon accumulations or deposits to the surface, to form various types of anomalies. This migration changes, somehow, the physical and chemical characteristics of the rocks and surface soil above the deposits or even the entire column of rock above them. The techniques that measure these changes are called indirect geochemical (and geophysical) detection methods (IDG) for two reasons: (1) those changes are an indirect or secondary result of hydrocarbon microseepage, and (2) such changes may also be caused by accumulations of biogenic methane. Different IDG methods are known, including the radiometric method, redox potential, bitumen analyses (luminescence or fluorescence), soil "salt" (mainly carbonates) analyses, trace element (metals) analyses, geobotanical analyses, soil mineralogy, hydrochemical analyses, and magnetic susceptibility measurements. An IDG method unrelated to hydrocarbon microseepage, but allowing the hydrocarbon deposit diagnosis, is helium surveying.

According to Saunders et al. (1999), combinations of methods such as airborne microwave sensing and laboratory analyses of soil gas hydrocarbons, shallow-source aeromagnetic and soil magnetic susceptibility measurements, aerial and surface gamma-ray measurements, as well as geomorphology are used to find oil and gas deposits. Results suggest that the use of these methods can substantially increase the probability of wildcat success and reduce finding costs in selected geologic settings; however, these surface methods cannot reveal depth, size, or quality of reservoirs, or even if producible hydrocarbons will be found.

According to Fedikow et al. (2009) and (2010), surface geochemical exploration (SGE) methods have a long and varied history in their application for searching oil and gas deposits, focused on the collection and analysis of soil gases, and the integration of the chemical expression of hydrocarbon microseepage with geological and geophysical (seismic) databases. By the other side, the various hydrocarbon reservoirs represent "metal-sources" of variable geochemical characteristics, i.e., they contain different trace elements in varying amounts. SGE techniques, therefore, can also identify metals associated to hydrocarbon microseepage from the same reservoirs by measuring mobile metal ions in surface soils and, in such a way, discriminate "productive" versus "non-productive" seismic targets. It can also be used to delineate bypassed resources of oil and gas in exhausted oilfields, to evaluate seismic targets for their prospectivity or to pioneer areas for the presence of oil and gas potentials. The source might be any accumulation of metals in contrast to the surrounding rocks—such as oil and gas—within the reservoirs and coal. Regardless of the type of metal-enriched zone at depth metal ions within that zone will be mobilized to the surface under a variety of mechanisms including the vapor/gas phase transport by light hydrocarbons, Hg-vapor, He, carbon dioxide and, perhaps, other mechanisms. This is evidence for an ongoing and dynamic geochemical system. By measuring mobile metal ions in surface soils, and through additional related geochemical determinations, it can document the presence of focused apical responses (anomalies) directly over the source regions, fact availed by many research, and case studies over known metal-/hydrocarbon-bearing zones.

Mac Elvain (1969) was the first who proposed the formation of hydrocarbon microseepage anomalies from the vertical ascent of ultra-small (colloidal size) gas bubbles through microfracture systems under water table above hydrocarbon deposits. The small sizes of such gas bubbles allow frictional effects to be overcome by the Brownian motion.

According to Klusman and Saeed (1996), three mechanisms are proposed for vertical migration of light hydrocarbons: diffusion, transport in aqueous solution, and buoyancy of microbubbles. Diffusion as a mechanism for primary migration of hydrocarbons from source rocks and as a transport mechanism in the near-surface unsaturated zone has been demonstrated. However, it fails to explain the rapid disappearance of surface anomalies after production from a reservoir begins. As a vertical migration mechanism, it also cannot account for the resolution observed in surface anomalies. Transport with water, either in solution or as a separate hydrocarbon phase, is important in secondary migration. However, computer modeling of the process fails to explain the observed resolution and rapid disappearance of surface anomalies. Authors, definitively, favor the vertical migration mechanism of buoyancy of microbubbles since computer modeling of this mechanism does explain surface observations: the close correspondence of surface anomalies to the surface projections of a reservoir, and the rapidly disappearing surface anomalies after the start of production.

Regarding chimneys, Price (1985) noted that these are the rock columns above hydrocarbon deposits modified by hydrocarbon vertically migration or by some other association of reduced species. Because of these modifications, the rocks of these chimneys are viewed to take on different lithological properties as compared to the surrounding country rocks off structure at equivalent depths. Although Pirson (1964), apparently, was the first to use the term "chimney," the concept can be traced further back. The creation of mineralized chimneys over hydrocarbon deposits has largely been attributed to hydrocarbon being "chemically reduced compounds." By this thinking, as they migrate into or through an environment, they "oxidize" to create a reducing environment, which leads to substantial changes in the mineralogy of the host sediments. The main product of microbial hydrocarbon oxidation is CO_2, which associates to water to form carbonate–bicarbonate species, which drastically change the pH of the system. The most apparent mineralogical change from a significant increase in the CO_2 concentrations is the precipitation of various carbonates, particularly calcite. The acidic solutions from the high CO_2 concentrations also alter or decompose clays, from neutralization of these solutions, which results in increased concentrations of silica and alumina. Such clay alteration is most probably responsible for the radiation lows reported in surface sediments over oil fields. Clay decomposition or alteration would liberate potassium from clays, including potassium-40, which, for many authors, is by far the major contributor to soil radioactivity. If this element is transported entirely from the system, a "hole" or low in the normal background readings would result. If the element is transported to the edges of the vertical projection of the HC deposit to the Earth's surface, where they precipitated from solution, a "halo" or "doughnut" of high values surrounding the low values over the surface projection of the hydrocarbon deposit would result.

Regarding anomalous distributions of iron and manganese in surface sediments over hydrocarbon deposits, Donovan and Dalziel (1977) noted that the passage of hydrocarbons and associated compounds such as hydrogen sulfide through surface rocks causes a reducing environment which, in turn, reduces iron and manganese to lower valence states, resulting in mobilization and removal of the elements, either precipitation of magnetite/maghemite or coprecipitation of iron and/or manganese with calcite in carbonate cements over hydrocarbon deposits. Magnetite formations, as well as other iron–manganese anomalies, form the basis for a number of different geophysical exploration techniques including soil magnetic susceptibility, aeromagnetic, and induced polarization.

Regarding microseepage from biogenic gas accumulations, Price (1985) remarked that it results into various anomalies such as magnetic, soil carbonate (including calcite-occluded CO_2), and radiation lows identical to anomalies created by microseepages from a thermogenic hydrocarbon deposit. Drilling such an anomaly would fail to result in a commercial oil discovery. Hence, explorationists should always be aware of the possibility of false anomalies caused by biogenic methane deposits on the normally oil-bearing stable shelves of sedimentary basins. Because of this possibility, only direct geochemical detection (DGD) methods which measure C_2–C_5 hydrocarbon concentrations, either relatively or absolutely, are specific for thermogenic hydrocarbon deposits.

3.5 Summary on Microbial Activity Related to Buried Mineralization and Hydrocarbon Deposits

Regarding microorganisms and metal mobility and transport, works on both porous media and aquatic sediments have investigated those processes whereby microbes, gases, and charge gradients can cause metals and elements to migrate through various media in the subsurface (Edwards et al. 2000). For example, a number of microbial associations have been implicated in the generation of gases above ore buried deposits (sulfur-reducing bacteria (SRB), dissimilatory iron-reducing bacteria (DIRB), methanogens, and methanotrophs). Therefore, beyond electrochemical and diffusive mobilization of metals above mineral deposits, it is necessary to begin examining the effect of rising gas bubbles on element mobility (Piotrowicz et al. 1979).

According to Price (1985) and Saunders et al. (1999), the interaction of bacteria with vertically migrating hydrocarbon constitutes the basis of most SGE methods, explaining many of the characteristics of hydrocarbon microseepage: the radiation lows and "halo" anomalies, the "disappearing anomalies," chimneys, iron/manganese anomalies, and others, which they are revealed by detectable modifications in geomorphic, seismic, magnetic, and radiometric properties. All hydrocarbon-oxidizing bacteria (found at all depths in the sediment column above hydrocarbon deposits, and in the edge waters of oil deposits) produce carbon dioxide and organic acids as a result of hydrocarbon oxidation. In addition, the sulfate reducers produce hydrogen sulfide and the denitrifiers produce free nitrogen and nitrous oxide. Microbial hydrocarbon oxidation during migration into or through an environment, creating a reducing one, consumes either free oxygen (O_2), or chemically oxygen complexes (SO_4^{-2} or NO_3^{-2}). In any case, the redox potential (eH) of the system is significantly affected. Being the main product of microbial hydrocarbon oxidation CO_2, it associates with water to form carbonate–bicarbonate species which drastically change the pH of the system. In addition to

CO_2, sulfur-reducing bacteria also produce H_2S, which also significantly affects the pH of the system. Such pH/eH changes result in totally new mineral stability fields: some minerals become unstable and are dissolved and mobilized; others are precipitated from solution.

3.6   Final Considerations and Conclusion

Many facts are common for the processes of metal mobilization, vertical transport to the surface, and the resultant accumulation of these vertically transported metals in surficial media from buried ore bodies and hydrocarbon deposits:

- Long-term water–rock reactions and microbial activity participating in metal source oxidation, releasing gases and ions with a chemically reduced character.
- A 'reducing chimney', reaching the surface, is formed above ore and hydrocarbon deposits from the vertical ascent of ultra-small (colloidal size) reduced gas bubbles (with reduced ions attached) through microfracture systems in the saturated zone, being the cause of the electrochemical effects suggested by the correspondent theory. As a consequence, rock columns above metal sources are somehow modified in their physical and chemical properties, as compared to the surrounding country rocks at equivalent depths. The most apparent mineralogical changes, due the presence of carbon dioxide and hydrogen sulfide, are the precipitation of various carbonates, particularly calcite, and decomposes of clays, which results into increased concentrations of silica and alumina ('acidic caps', in the unsaturated zone). These changes are responsible for radiation lows (due liberation of potassium), and magnetic susceptibility/chargeability increases (caused by either the precipitation of magnetite/maghemite, or the coprecipitation of iron and/or manganese with calcite in carbonate cements over the metal sources).
- For the unsaturated zone, with a more fracture and porous character, a gas advective transport mechanism as barometric pumping, as well as capillary rise support moving upward ions and submicron metal particulates to the surface.
- Redistribution of ions in the near-surface environment by downward-percolating groundwater (after rainfall) as well as by the upward effects of evaporation and capillary rise explain soil metal accumulations in a very shallow (10–30 cm) 'metal accretion zone'.

As conclusion, it is outlined the general features that characterize the processes of metal mobilization, vertical transport to the surface, and the resultant accumulation of these vertically transported metals in surficial media from buried ore bodies and hydrocarbon deposits: microbial activity and water–rock reactions with gas (hydrocarbons, N, CO_2, H_2, and others) generation during target oxidation; ascending reduced gas microbubbles (colloidal size) with reduced metal ions attached, which results in 'reducing chimneys' reaching the surface; barometric pumping and capillary rise moving upward ions and submicron metal particulates into the unsaturated zone; redistribution of ions in the near-surface environment by downward-percolating groundwater (after rainfall), as well as by the upward effects of evaporation and capillary rise, all of which explain the soil metal accumulations in a very shallow (10–30 cm) 'metal accretion zone'.

References

Birell R (1996) MMI geochemistry: mapping the depths. Min Mag 174(5):306–307

Cameron EM, Hamilton SM, Leybourne MI, Hall GEM, McClenaghan B (2004) Finding deeply-buried deposits using geochemistry. Geochem Explor Environ Anal 4:7–32

Donovan TJ, Dalziel MC (1977) Late diagenetic indicators of buried oil and gas: U.S. Geological survey open-file report, 77–817, 44 pp

Edwards KJ, Bond PL, Druschel GK, McGuire MM, Hamers RJ, Banfield JF (2000) Geochemical and biological aspects of sulphide mineral dissolution: lessons from iron mountain, California. Chem Geol 169:383–397

Fabris AJ, Keeling JL (2006) Exploration through transported cover: summary of approaches and methods. Regolith 2006 - consolidation and dispersion of ideas. CRC LEME, Perth

Fedikow MA, Bezys RK, Nicolas MB, Prince P (2009) Preliminary results of soil geochemistry surveys in support of shallow gas exploration, Manitou area, Manitoba (NTS 62G2). In: Report of activities 2009, manitoba innovation, energy and mines, manitoba geological survey, pp 193–206

Fedikow MA, Bezys RK, Nicolas MB, Prince P (2010) Ligand-based partial extraction of near-surface soil samples: an innovative geochemical approach to shallow gas exploration. Southwestern Manitoba. GeoCanada 2010 – Working with the Earth

Goldberg IS (1998) Vertical migration of elements from mineral deposits. J Geochem Explor 61:191–202

Govett GJ (1973) Differential secondary dispersion in transported soils and post-mineralization rocks; an electrochemical interpretation. In: JONES MJ (ed) Geochemical Exploration 1972. Institute of Mining and Metallurgy, London, pp 81–91

Hall G, Hamilton SM, McClenaghan MB, Cameron EM (2004) Secondary Geochemical Signatures in Glaciated Terrain: SEG 2004, Perth

Hamilton SM (1998) Electrochemical mass-transport in overburden: a new model to account for the formation of selective leach geochemical anomalies in glacial terrain. J Geochem Explor 63:155–172

Hamilton SM (2000) Spontaneous Potentials and Electrochemical Cells: In: Geochemical remote sensing of the sub-surface, Hale, vol 7. Elsevier, Amsterdam, pp 421–426

Hamilton SM (2007) A prospector's guide to the use of selective leach and other deep penetrating geochemical techniques in mineral exploration; Ontario geological survey, Open File Report 6209, 39 pp

Klusman RW (2009) Transport of ultratrace reduced gases and particulate, near-surface oxidation, metal deposition and adsorption. Geochem Explor Environ Anal 9:203–213

Klusman RW, Saeed MA (1996) Comparison of light hydrocarbon microseepage mechanisms. In: Schumacher D, Abrams MA (eds) Hydrocarbon migration and its near-surface expression: AAPG Memoir 66, p 157–168

Mac Elvain R (1969) Mechanics of gaseous ascension through a sedimentary column. In: Heroy WB (ed) Unconventional methods in exploration for petroleum and natural gas. Southern Methodist University Press, Dallas, p 15–28

Malmqvist L, Kristiansson K (1984) Experimental evidence for an ascending microflow of geogas in the ground. Earth Planet Sci Lett 70:407–416

Mann AW, Birrell RD, Fedikow MAF, de Souza HAF (2005) Vertical ionic migration: soil anomalies, and sampling depth for mineral exploration. Geochem Explor Environ Anal 5:201–210

Mann AW, Gay LM, Birrell RD, Webster JG, Brown KL, Mann AT, Humphreys DB, Perdrix JL (1995) Mechanism of formation of mobile metal ion anomalies. In: Report 153. Minerals and Energy Research Institute of Western Australia

Mann AW, Mann AT, Humphreys DB, Dowling SE, Staltari S, Myers L (1997) Soil geochemical anomalies - their dynamic nature and interpretation. In: Report 184, vol 1. Minerals and Energy Research Institute of Western Australia

Pardo M, Rodríguez O (2009) El *Complejo Redox*. Consideraciones Metodológicas y Empírico-Teóricas. Tercera Convención Cubana de Ciencias de la Tierra GEOCIENCIAS 2009. Memorias. La Habana: Centro Nacional de Información Geológica. ISBN 978-959-7117-19-3

Pardo M, Stout R et al (2003) El *Complejo Redox*: Estado del Arte en el 2002. Memorias del V Congreso Cubano de Geología, La Habana

Piotrowicz SR, Duce RA, Fasching JL, Weisel CP (1979) Bursting Bubbles and Their Effect on the Sea-to-Air Transport of Fe, Cu and Zn. Marine Chemistry 7:307–324

Pirson SJ (1964) Projective well log interpretation, Part 1: World Oil, August, pp 68–72

Price LC (1985) A critical overview of and proposed working model for hydrocarbon microseepage. In: U. S. Department of the Interior Geological Survey. Open-File Report, pp 85–271

Sato M, Mooney HM (1960) The electrochemical mechanism of sulphide self potentials. Geophysics 25:226–249

Saunders DF, Burson KR, Thompson CK (1999) Model for hydrocarbon microseepage and related near-surface alterations. AAPG Bulletin 83(1):170–185

Smee BW (1998) A new theory to explain the formation of soil geochemical responses over deeply covered gold mineralization in arid environments. J Geochem Explor 61:149–172

Sutherland D (2011) Soil gas hydrocarbons: an organic geochemistry that detects hydrocarbon signatures in surficial samples to locate and identify deeply buried targets. In: Search and discovery article #40684

Chapter 4
The *Redox System*: The *Redox Complex* Database and Interpretation System

Abstract Along its development, the *Redox Complex* (ORP, magnetic susceptibility, spectral reflectance, and soil geochemistry) has been needed of tools for the storage, report, and interpretation of their data. For this reason, it was designed a database and application system to solve in a quick and reliable way, all the storage processes, reports, graphics, and the corresponding interpretations. To carry out the design of the different kinds of applications which responded to the qualitative and quantitative data interpretations, it was necessary to model mathematically the expert's experience. First was designed all the working algorithms with respect to the empirical–practical experiences of the different interpretive steps of the *Redox Complex*. For this objective, the methods of the engineering of the knowledge were used. For the design of the *Redox System* the techniques of the UML (Unified Model Language) diagrams were used. The *Redox System* transcends the limits of a simple calculation and storage program, which it can be used by other specialists to obtain results of the interpretation without being expert in the topic. A user manual of the *Redox System* was drawn up and it was gotten the Certification of Facultative Legal Deposit of Protected Works.

Keywords Redox Complex · ORP · Magnetic susceptibility · Soil geochemistry · Unified Model Language · Engineering of the knowledge

4.1 Introduction

It was designed a database and application system (Booch G 1996; Booch G et al. 1999a, b) to solve in a quick and reliable way for the storage, report, and interpretation of data corresponding to the *Redox Complex* (ORP, magnetic susceptibility, spectral reflectance, and soil geochemistry). To carry out the design of the different kinds of applications which responded to the qualitative and quantitative interpretations of the data, it was necessary to model mathematically the expert's experience. First was designed all the working algorithms with respect to the empirical–practical

© The Author(s) 2016
M.E. Pardo Echarte and O. Rodríguez Morán, *Unconventional Methods for Oil & Gas Exploration in Cuba*, SpringerBriefs in Earth System Sciences, DOI 10.1007/978-3-319-28017-2_4

experiences of the different interpretive steps of the *Redox Complex*. The *Redox System* transcends the limits of a simple calculation and storage program, which it can be used by other specialists to obtain results of the interpretation without being expert in the topic. A user manual of the *Redox System* was drawn up and it was gotten the Certification of Facultative Legal Deposit of Protected Works.

4.2   Materials and Methods

It used the methods of the engineering of knowledge (Larman 2002) to model the referred algorithms. For the design of the *Redox System* the techniques of the UML (Unified Model Language) diagrams were used.

The information flow generated by the *Redox System* is shown in Fig. 4.1. It starts from the observed data, which are introduced in the database by means of a formulary (Data Catcher). It also takes into account the transformations of reduction/normalization, determination of the background level of the observations, the generation of the corresponding graphs, and reports of sector and line.

The basic entities and their relationships are shown in Fig. 4.2. This figure shows the relationship of *one to many* of each of these entities, which run from the general

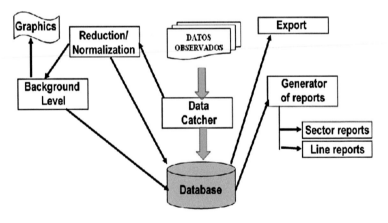

Fig. 4.1 General information flow in the *Redox System*

Fig. 4.2 Relationship of *one to many* of each of these fundamental entities of the system

to the particular, e.g., Project, Objective, Sector, Line (or Profile), and Stations (Stakes).

At the database system that supports the *Redox System* not only the observed data necessary for the interpretation process, but also the descriptive data of the main attributes are stored; these are:

- The Geospatial information (Lambert coordinates, geographical coordinates, name and number of the cartographic sheet at 1:50000 scale).
- Different kinds of data (observations) reflected in the field notebook.
- Identification of the soil sample taken to be sent to the laboratory.
- Chemical elements determined in the laboratory.

In addition to these explanatory elements of the data, the observations of the *Redox Complex* are included: Magnetic susceptibility, ORP, spectral reflectance, and content of chemical elements determined in the observation point.

4.3   Results

4.3.1   Redox System *Environment*

The *Redox System* environment (Fig. 4.3) consists of the facilities provided by the handler and the wizard interface designs or forms of Microsoft Access (Reenskaug et al. 1996; Riel 1966).

There is a defined folder structure for *Redox System*. On a folder named ComplejoRedox System is structured. Inside this folder there are the following other folders: Aplicaciones, Ayuda, Imagenes, ModelosMatematicos y Reportes (Fig. 4.4).

Fig. 4.3 Main *Redox System* interface

Fig. 4.4 Folder structure
defined for the *Redox System*

◢ 📁 ComplejoRedox

  ▷ 📁 Aplicaciones

  📁 Ayuda

  ▷ 📁 Imágenes

  📁 ManualUsuario

  ▷ 📁 ModelosMatematicos

  ▷ 📁 Reportes

- Database (SistemaRedox_v_1_9.accdb) is located directly in the **ComplejoRedox** folder.
- In the **Aplicaciones** folder the programs developed in Visual Basic that are eventually used by the Redox System are stored.
- In the **Ayuda** folder pdf files are stored on which the *Redox System* aid is designed.
- The **Imagenes** folder has within a number of folders references to images that describe the observations for each project undertaken.
- The **ModelosMatematicos** folder contains the Excel objects over which the whole process of mathematical modeling of the interpretation process is validated. These Excel objects are automatically linked to the interface of interpretation of *Redox System*.
- The **Reportes** folder automatically stores all items generated as reports by the *Redox System*.

The system is accessed by selecting the name in a list of the authorized user, the private key of the user is typing. This key should govern in a particular domain of the system.

The menu includes the following main entrances: Autorización/Captador/ Reportes/Interpretación/Herramientas/Exportar/Importar/Nomencladores/Creditos/ Ayuda/Salida. For each of the elements of the first line of the menu, there is usually a series of submenu that offers a range of variety of prosecutions of the system.

4.4   Interface Environment

Each of the interfaces of the *Redox System* will be explained below, noting, in this way, its main use:

- **Authorization Module Interface**. Authorization Module system is composed of Authorization User Interface. This interface is manipulated only by the Administrator.

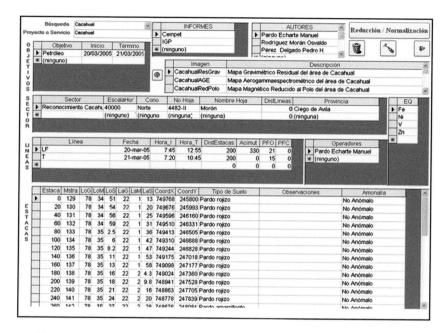

Fig. 4.5 Capture Module Interface of the *Redox System*

- **Capture Module Interface.** This module is one of the most important of *Redox System* (Fig. 4.5). Here is introducing, using the keyboard, data measured in the work of *Redox Complex*. The spatial distribution of the information of the interface is in accordance with the basic and hierarchical structure.
- **Primary Elaboration Module Interface.** This module is very important because it has one of the elements essential to modeling the general process of interpretation: the estimated background level of the *Redox Complex* attributes and the corresponding Reduction/normalization of the data (Fig. 4.6).
- **Report Module Interface.** Generating reports is framed in two variants, Sector Reports and Line Reports. An example of an interpreted Line Report is shown in Fig. 4.7.
- **Interpretation Module Interface.** The interface consists of a link to an Excel object, which is carrying the methodology of mathematical modeling of the interpretation process for purposes of oil, minerals, environment pollution, and metal archeological burials (Fig. 4.8).
- **Tools Module Interface.** In this part of the menu programs that have been helpful in the system are concentrated. These tools are coordinate transformation, spatial zoning, cluster analysis, generation of curves, and cartographic trapezoids.

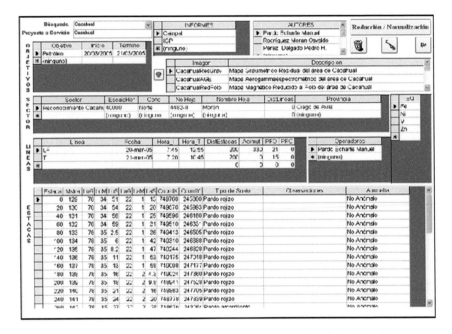

Fig. 4.6 Primary Elaboration Module Interface of the observed data of the *Redox System*

- **Export Module Interface**. This interface establishes the communication of *Redox System* with commercial systems. There are actions that the *Redox System* does not hold, since there are commercial programs that do, for example, mapping, among others. Systems to which the *Redox System* established relationship by export files are AutoCad (dxf), GeoSoft (xyz), and Surfer (dat and bln).
- **Nomenclature Interface**. The *Redox System* established a series of codifiers which can only be updated and augmented by the System Administrator.
- **Help Interface**. The aid was made into a pdf file with a list of hyperlinks that provide knowledge about different parts of the system to be elucidated. In Fig. 4.9, the start of the *Redox System* Help is shown.

A *Redox System* user manual which replaces the Help of the system was drawn up. In addition, the Facultative Legal Deposit Certification of Protected Works was obtained with the Registration No. 1589-2005 (Rodríguez 2005).

In Table 4.1, the volume of information on the *Redox System* for each of the areas of application to date of writing of this paper is shown. In Fig. 4.10 the corresponding pie chart shows this table.

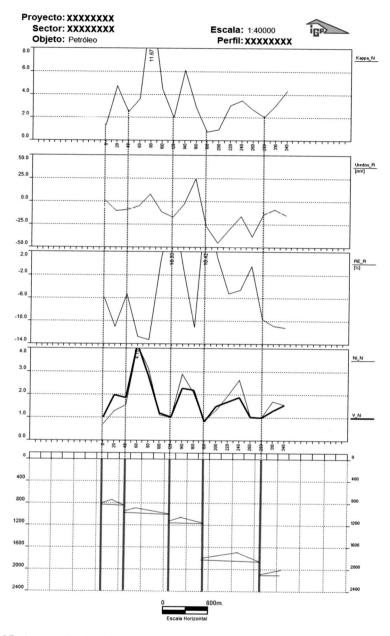

Fig. 4.7 An example of an interpreted profile by the *Redox System* over an oil target

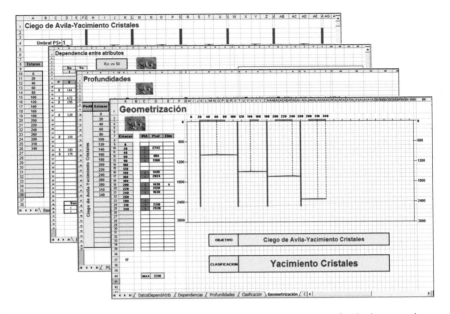

Fig. 4.8   Some examples of Excel sheets which carrier the methodology for the interpretation

Fig. 4.9   The beginning of the *Redox System* Help

Table 4.1 Information volume of the *Redox System* for each of the application spheres

Application sphere	Number of observations	Total volume (km)
Oil and Gas	949	204189
Metallic mineral	1873	41743
Environmental studies	318	7635
Archeological studies	50	26
Total	3190	253593

Fig. 4.10 Percentage ratio chart of the information contained in the *Redox System* for each of the areas of *Redox Complex* applications

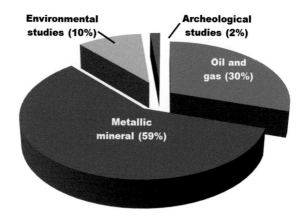

4.5 Conclusion

With the design and operation of the *Redox System* (database and applications), you can give quick and reliable response to storage processes, reports, graphs, and interpretations relating to the application and development of *Redox Complex*.

References

Booch G (1996) Análisis y Diseño Orientado a Objetos con aplicaciones, 2ª edición, Addison-Wesley/Diaz de Santos. Disponible en: http://www.cuspide.com/detalle_libro.php?isbn=9684443528

Booch G et al (1999a) Unified Modeling Language UML. Notación guide and UML Semantics, version, 1.3. Disponible en: http://www.rational.com

Booch G et al (1999b) El lenguaje unificado de modelado. Disponible en: http://www.agapea.com/El-lenguae-unificado-de-modelado-n11514i.htm

Larman C (2002) UML y Patrones: Una introducción al análisis y diseño orientado a objetos y al proceso unificado, 2da, Prentice-Hall. Disponible en: http://www.fitco.edu.co/iacademica/sistemas/Agora/articulos/UML.pdf

Reenskaug T et al (1996) Working with objects, manning. Disponible en: http://www.2020ok.
    com/4013.htm
Riel A (1966) Object-oriented design heuristics. Disponible en: http://www.iam.unibe.ch/~girba/
    download/Mari05aSAIL.pdf
Rodríguez O (2005) Manual de Usuario *Sistema Redox*. Derecho de Autor. Certificación de
    Depósito Legal Facultativo de Obras Protegidas. CENDA. Registro: 1589–2005

Appendix

Illustrations of some *Redox Complex* Applications to Oil & Gas Exploration in Cuba

The illustrations of some of the applications of geophysical-geochemical unconventional methods for oil exploration in Cuba consider the regions of Habana-Matanzas (Varadero Oil Field, Cantel Oil Field and Madruga Prospect) and Ciego de Ávila (Pina Oil Field, Cristales Oil Field, Jatibonico Oil Field, Jatibonico Oeste Prospect and Cacahual Prospect).

The methods considered contemplated, in some cases, airborne gamma spectrometry (K/eTh ratio, dimensionless) and reduced Redox Potential (ORP) in mV and, in others, the *Redox Complex* with reduced or standard attributes (Magnetic Susceptibility in 10^{-3} SI, ORP in mV, Spectral Reflectance in % and Content of Chemical Elements (Ni or V, or Pb) in ppm; as that order, from the top to the bottom of the figure).

In all cases, the anomalous complex of interest corresponds to the correlation of minimum K/eTh ratio, minimum ORP and, in the case of *Redox Complex*, Magnetic Susceptibility highs with ORP lows, Spectral Reflectance lows and maximum Content of Chemical Elements.

In some cases, the illustrations are accompanied with producer horizon, as seismic data, or *Redox Complex* data interpretation; in others, only the range shown Anomalous Response.

© The Author(s) 2016
M.E. Pardo Echarte and O. Rodríguez Morán, *Unconventional Methods for Oil & Gas Exploration in Cuba*, SpringerBriefs in Earth System Sciences, DOI 10.1007/978-3-319-28017-2

SECTOR VARADERO - CANTEL

PERFIL VARADERO
ESCALA 1: 40 000

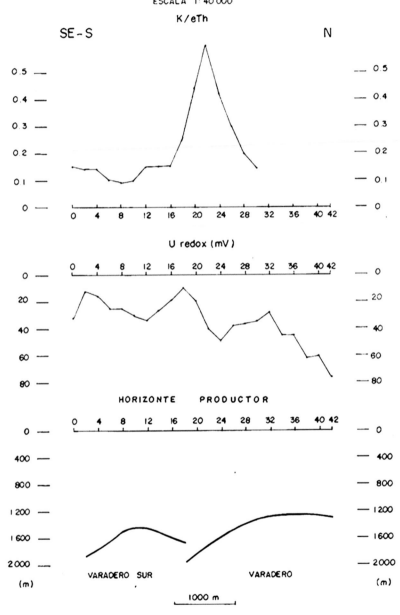

K/eTh

SECTOR VARADERO - CANTEL

PERFIL CANTEL

ESCALA 1:40 000

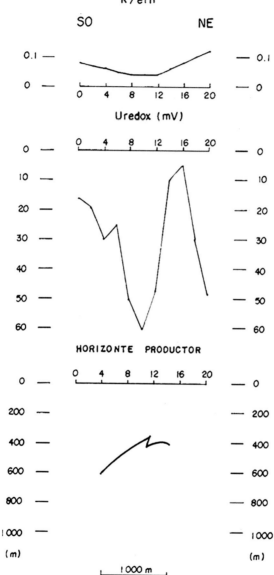

K/eTh

HORIZONTE PRODUCTOR

1000 m

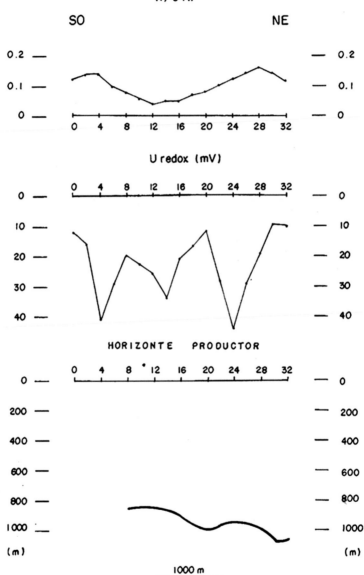

SECTOR PINA
PERFIL PINA I
ESCALA 1:40 000
K / e Th

SECTOR MADRUGA

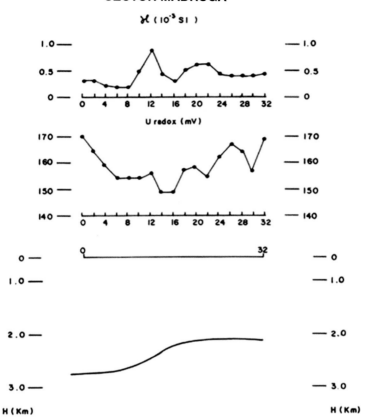

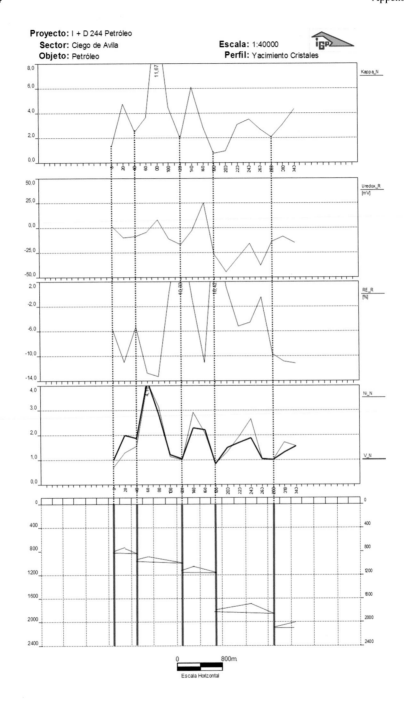

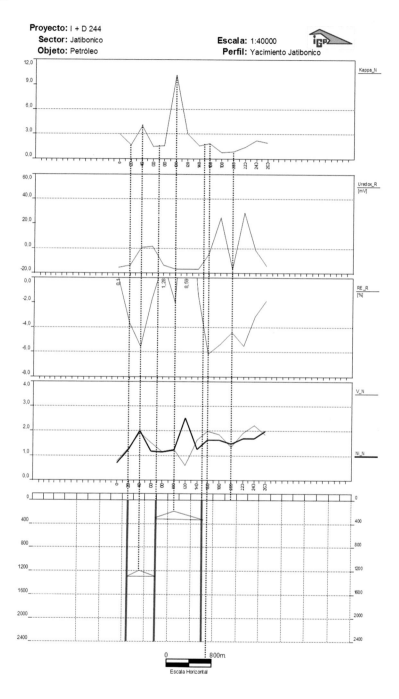

Proyecto: I + D 227
Sector: Pina
Objeto: Petróleo

Escala: 1:40000
Perfil: Pina Este

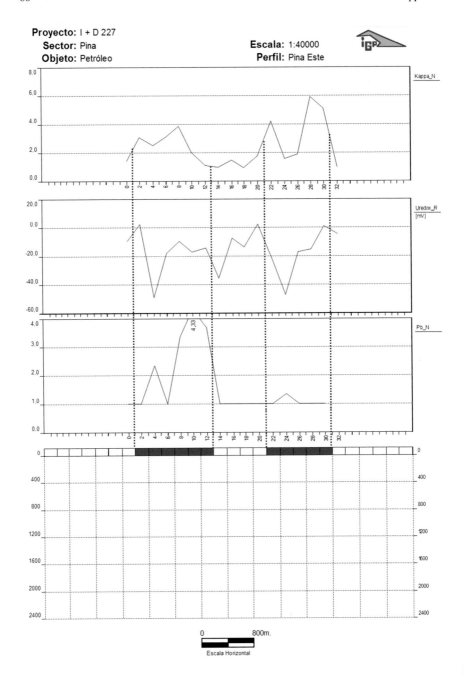

0   800m.
Escala Horizontal

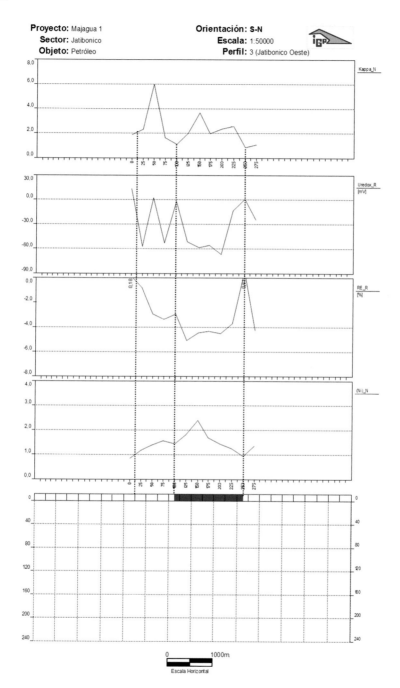

Proyecto: Majagua 1   Orientación: S-N
Sector: Jatibonico   Escala: 1:50000
Objeto: Petróleo   Perfil: 3 (Jatibonico Oeste)

0   1000m
Escala Horizontal

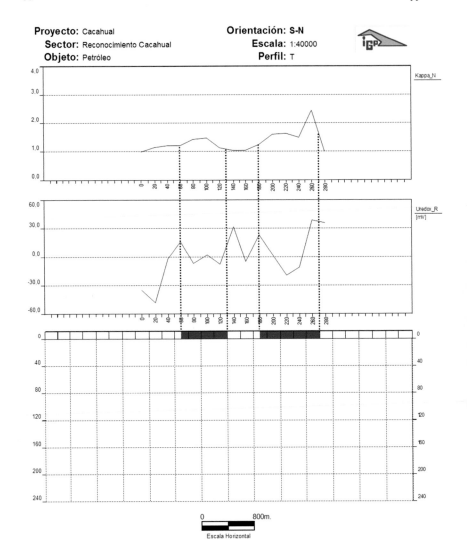

Proyecto: Cacahual   Orientación: S-N
Sector: Reconocimiento Cacahual   Escala: 1:40000
Objeto: Petróleo   Perfil: T

Kappa_N

Uredox_R [mV]

0   800m.
Escala Horizontal

Index

A
Archeology, 10, 11, 14, 24, 25

B
Buried ore bodies, 21, 38, 45

C
Chimneys, 4, 37, 38, 40, 43–45

E
Engineering of the knowledge, 49, 50
Environment, 3, 8, 10, 11, 14, 19–22, 24, 25, 34, 37, 41, 43–46, 51, 52

G
Geochemical exploration, 4, 15, 37, 38
Geochemistry, 39
Geological prospecting, 1, 15, 19–21, 23, 34
Geophysics, 37, 38, 40

H
Hydrocarbon deposits, 21, 22, 37, 38, 40–45

L
Language, 49, 50

M
Magnetic susceptibility, 2, 7, 8, 12, 19–25, 28, 29, 31–34, 39, 40, 42, 45, 49, 51, 59

Metallic mineral occurrences, 9
Metallic sources, 2, 13, 15, 38, 42, 45
Microbial activity, 37, 44, 45
Mobile metal ions (MMI), 2, 3, 15, 38, 39, 42

O
Oil-gas deposits, 42
Ore deposits, 41
ORP, 4, 7, 8, 10, 11, 19–22, 25, 26, 28, 30–34, 49, 51, 59

R
Redox Complex, 1, 4–14, 15, 19–21, 24–27, 34, 37, 39, 40, 49, 53, 57, 59
Reduced, 4, 8–10, 12, 13, 20, 21, 30–34, 37, 38, 41, 43, 45, 59
Resource estimate, 1

S
Soil geochemistry, 2, 19–22, 24, 34, 37, 39, 40, 49
Soil redox potential, 2, 6, 20

U
Unconventional exploration techniques, 19, 34
Unconventional geophysics-geochemical exploration, 1, 15
Unified Model, 49, 50

© The Author(s) 2016
M.E. Pardo Echarte and O. Rodríguez Morán, *Unconventional Methods for Oil & Gas Exploration in Cuba*, SpringerBriefs in Earth System Sciences, DOI 10.1007/978-3-319-28017-2

Printed in the United States
By Bookmasters